Wallace C. Sabine

COLLECTED PAPERS

ON ACOUSTICS

by

WALLACE CLEMENT SABINE

*Late Hollis Professor of Mathematics and Natural Philosophy
in Harvard University*

76272

With a New Introduction by

FREDERICK V. HUNT

*Rumford Professor of Physics and
Gordon McKay Professor of Applied Physics
Harvard University*

DOVER PUBLICATIONS INC., NEW YORK

This Dover edition, first published in 1964, is an unabridged and unaltered republication of the work first published by the Harvard University Press in 1922, to which has been added a new Introduction by Frederick V. Hunt.

Library of Congress Catalog Card Number: 64–18864

Manufactured in the United States of America

Dover Publications, Inc.
180 Varick Street
New York 14, N.Y.

INTRODUCTION TO DOVER EDITION

In the short span of twenty years, extending from 1895 to 1915, Wallace Clement Sabine founded and brought to a high degree of maturity the science of architectural acoustics. This fact alone would be sufficient to endow these *Collected Papers* with historical significance, but they have more to offer. It is seldom that the personal character and quality of an author are so clearly revealed as in these writings. Sabine emerges as a gracious and dedicated man, at once reserved and intense, and as a meticulous experimenter guided by the highest ideals of scientific integrity. In modern terminology Sabine would be characterized as an applied physicist; he never wavered in his devotion to the achievement of practical and useful results. To this end he published most of these papers in architectural and building-trade periodicals, rather than in the learned-society journals, because he felt that in this way he could better reach the people who were most desperately in need of the help he had to offer.

Wallace Clement Sabine, the son of Hylas Sabine and the grandson of John Fletcher Sabine, was born in Richwood, Ohio, on 13 June 1868. After graduating from Ohio State University, B.A. in Physics, 1886, he continued with graduate studies at Harvard University, to which he had been irresistibly attracted by a chance encounter with Professor John Trowbridge in 1884. His connection with Harvard remained unbroken thereafter until his untimely death following an operation for a liver infection on 10 January 1919. He was awarded a Harvard Master of Arts degree in 1888 but he never offered himself as a candidate for the Ph.D. because, as Sabine himself explained later, "when the proper time came for me to do so, I should have been my own examiner!" This omission was rectified in due course when first Brown University (1907) and then Harvard (1914) awarded him honorary doctorates in science. Two years of service as research assistant to Professor Trowbridge and five years as an instructor paved the way for Sabine's promotion to the assistant professorship with which this account of the origins of architectural acoustics begins.

As Sabine points out in the first paper of this collection, he under-

took these investigations not by choice but at the request of President
Eliot to "do something" about the acoustical difficulties in the lecture
room of Harvard's new Fogg Art Museum,* which had just been completed in 1895. The assignment of this chore to Sabine was at least in
part an accident of timing. His previous researches in collaboration
with Professor Trowbridge had been concerned with optics and electricity and had revealed no special predilection for the problems of
acoustics. As a newly appointed and relatively uncommitted assistant
professor, however, Sabine's insight and zeal had been under recent
review by the University administration. So the challenge was laid
down. President Eliot would have been satisfied with even a partial
correction of the acoustics if this would have made the new Fogg lecture room usable, but Sabine chose to accept the challenge in a broader
sense. He quickly widened the scope of his studies to include nearby
Sanders Theatre (which had excellent acoustics), the large lecture
room in the Jefferson Physical Laboratory (which had tolerable acoustics and the prime advantage of accessibility), and the constant-temperature room in the sub-basement of Jefferson (which became his
reverberation chamber). With a few organ pipes, a stop watch, and
his own well-disciplined sense of hearing, Sabine turned these rooms
into laboratory tools for a basic investigation of the behavior of
sound in enclosed spaces.

In common with many of the memorable men of science, Sabine had
the rare ability to regard a complex physical situation and to pick out
the key problem whose solution would open the door to understanding. Along with this ability to see the real problem, he had the equally
rare knack of being able to characterize it articulately. Almost at the
outset, therefore, we find Sabine expressing with objective clarity the
necessary and sufficient conditions for good hearing. These simple
statements, that simultaneous sounds should preserve their frequency
balance and that successive sounds should remain clear and distinct
from each other and from extraneous sounds, are just as true and comprehensive today as they were when Sabine wrote them in 1898.
Modern technology and materials have given the designer new free-

* The name of this building was subsequently changed to Hunt Hall, in honor of its
architect, Richard Morris Hunt, when a still newer Fogg Art Museum was erected on
the University campus in 1928.

doms of execution in achieving frequency balance, reverberation control, and sound isolation, but the designer's goals are still the same.

In spite of their apparent simplicity, these declarations of acoustical intent were wholly novel and without precedent. As Sabine had quickly discovered, architectural acoustics had been the most neglected branch of the science of sound and only a handful of mid-nineteenth-century authors had even attempted to deal with it. The French architect, T. Lachéz, brought out *L'acoustique et l'optique des salles de réunions* in 1848, and a revised edition of it was called for by 1879. Dr. J. B. Upham, a Boston physician, published a series of papers in 1853 in which he considered reverberation and showed that hanging curtains could reduce it. Joseph Henry, who is better known for his discovery of self-induction, inserted a discussion of "Acoustics Applied to Public Buildings" in the 1856 *Annual Report* of the Smithsonian Institution. Henry had determined experimentally the smallest time delay that would allow returning sound to be perceived as a distinct echo, and his remarks indicated that he had a firm grasp of the basic phenomena of sound reflection. The key word that characterizes all these discussions is "qualitative." This state of affairs was well summarized in 1864 by the British architect, T. Roger Smith, in *The Acoustics of Public Buildings*, a compact and now rare volume in Weale's series of "Rudimentary Treatises."

There is a remarkable similarity between these scattered nineteenth-century writings and those of the ancients. Lucretius described the phenomena of reverberation, without remedial prescription, in lyric verse: "One voice . . . disperses suddenly into many voices . . . some dashed upon solid places and then thrown back . . . . Therefore the whole place is filled with voices . . . all around boils and stirs with sound . . . ." The Roman architect, Vitruvius, whose only surviving monument is a ten-volume treatise on architecture, was the only one among either the ancients or the pre-Sabinites who came close to matching the clarity of Sabine's statement of the conditions for good hearing. The directions for achieving good amphitheatre acoustics that were laid down by Vitruvius have a ring of modern relevance, but only if one attaches to the words he used a modern interpretation that goes far beyond the level of understanding extant in his time. In the Aristotelian *Problemata*, the question is raised, "Why is it that when the orchestra of a theater is spread with straw, the chorus makes less sound?" Hardly more revealing than this question was the only answer that T. R. Smith could find two millennia later: "When there is too much resonance in a room, carpets or curtains may be advantageously employed to lessen it." Unfortunately, neither question nor answer went very far beyond the tacit acoustical advice given in Exodus XXVI, where the horizontal length of the curtains of goat's hair specified for the tabernacle insured that the curtains would hang in generous sound-absorbing folds.

The major acoustical problem besetting the Fogg lecture room was obviously excessive reverberation, but Sabine was not content either to judge this problem or to prescribe for its correction on merely qualitative grounds. His patient and painstaking efforts to put the problem and its solution on a quantitative basis, and his success in doing this, provide a unique opportunity to identify the exact time and place at which the *science* of architectural acoustics came into being. The history of science is spiced with many episodes that describe a moment of singular revelation. The most durable and appealing of these stories are almost surely apocryphal, such as the ones that describe Pythagoras listening to the consonant sounds of hammers of different weights striking an anvil, Archimedes shouting "Eureka!" and Galileo dropping stones of different sizes from the leaning tower of Pisa. Architectural acoustics had a similar moment of truth that is better documented. During the fall of 1898, Sabine was living with his mother in quarters on Garden Street in Cambridge. His recommendations for improving the Fogg lecture room had been carried out successfully, and President Eliot was urging him to accept the invitation of Major Higginson to give acoustical advice in connection with the planning of a new Boston Music Hall. Before accepting this important assignment, Sabine decided to review his work on the Fogg lecture room. We find him then, on Saturday evening, 29 October 1898, quietly poring over his data on the duration of residual sound as a function of the running length of the Sanders Theatre seat cushions he had brought into the Fogg lecture room. Suddenly he shouted triumphantly from his study: "Mother, it's a hyperbola!" This simple but penetrating observation, that the product of total absorption and the duration of residual sound was a constant, gave Sabine the quantitative hold he had been seeking on the phenomena of reverberation. Now he could write to President Eliot the next morning as follows: "When you spoke to me Friday in regard to Music Hall, I met the suggestion with a hesitancy the impression of which I now desire to correct. At the time I was floundering in a confusion of observations and results which last night resolved themselves in the clearest manner." After a brief description of this "clearest manner," Sabine concluded his letter on a new tone of confident assurance: "It is only necessary to collect further data in order to predict the character of any room that may be planned, at least as respects reverberation."*

With his hyperbola in hand and with his hesitancy corrected, Sabine threw himself into the planning of the new Music Hall with sustained enthusiasm. Even though he was sure that the character of a room "as respects reverberation" could be predicted quantitatively, Sabine realized, as modern practitioners do, that many other factors also contribute to the achievement of good hearing conditions. His judgment as a trained physicist was his chief guide in these matters. No detail of the interior construction escaped his attention and many of its features give evidence of his appreciation of their acoustical effect. He insisted, for example, on a modest ceiling height for control of reverberation by restriction of volume, on shallow balconies to avoid deep sound shadows, on a shallow stage house with angled walls and ceiling to conserve orchestral output, and on wall niches and a deeply coffered ceiling to promote sound scattering and diffusion. The new Boston Music Hall, now known as Symphony Hall, was formally opened on 15 October 1900 to the acclaim of musicians and critics alike, and it still ranks among the three or four best auditoriums in the world for symphonic music. As the first music hall to be built with rationally preplanned acoustics, it remains a fitting acoustical monument to its designer.

His success with the acoustics of Symphony Hall immediately established Sabine as the master of his subject, and his services as a

---

* This letter and some of the other biographical data used here have been extracted, with kind permission, from William Dana Orcutt's affectionate biography, *Wallace Clement Sabine: A Study in Achievement* (Plympton Press, Norwood, Mass., 1933). The incident of the hyperbola was described to me by Dr. Paul E. Sabine—a second cousin of W.C.S.—who said he got the story from W.C.S.'s family.

consultant were in constant demand throughout the rest of his life. He responded generously to such requests for help, often at considerable personal sacrifice, and he accepted fees only when forced to do so by grateful clients. Each new inquiry presented new challenges and he welcomed them all as opportunities to "collect further data." Even without the stimulus of these practical examples, Sabine knew that many of the questions he had dealt with needed more complete answers. For example: How did the sound absorption of various materials vary with frequency? How much precision was really needed in the control of reverberation? How could the distribution of sound be controlled? What were the acoustical effects of different arrangements for heating and ventilating? And how could better sound insulation be achieved? With dogged persistence Sabine pursued these questions in the field and in the laboratory. The results of this sustained assault on the problems of architectural acoustics are presented in these *Collected Papers*.

In two of his later papers, Sabine singled out for special attention "The Correction of Acoustical Difficulties" and "Theatre Acoustics." He had already learned what every acoustical consultant still learns, that a wholly disproportionate amount of effort must be devoted to the correction of acoustical troubles that could have been avoided in the first place by rational planning. By parading a series of examples of adroit acoustical management, Sabine showed that in remedial correction as in original design it is often possible to conserve both reverberation control and the features that promote a favorable distribution of sound. Sabine was one of the first to use spark photographs of sound waves in model rooms as a tool for studying the details of sound distribution, and many revealing pictures obtained in this way are exhibited in "Theatre Acoustics" and in the summary paper he contributed to the *Journal of the Franklin Institute* in 1915. This technique represents one of Sabine's efforts to put the distribution of sound, like reverberation, on a firm quantitative basis. Unfortunately, this goal eluded him as it has continued to elude all who have followed him.

Three of these *Collected Papers* represent quiet pools beside the main stream of development of architectural acoustics. In his paper on "Melody and the Origin of the Musical Scale," Sabine pointed out that

Helmholtz had overlooked the effects of reverberation in his classic treatise on *Sensations of Tone*. It became obvious, after Sabine explained it, that reverberation can convert the sequential tones of melody into the simultaneously heard tones of harmony, and that Helmholtz's elaborate rationale of harmony could, therefore, have been applied with equal validity to explain the origin of musical scales. On the basis of a slight extension of these ideas, Sabine ventured also to suggest somewhat diffidently that even the evolving forms of Christian worship may have been profoundly influenced by the acoustics of churches. Sabine did not regard his "Sense of Loudness" as worthy of any but local publication, but it takes on fresh interest as the account of a pioneer experimental measurement of what would now be called an equal-loudness contour. His ingenious adaptation of the techniques of reverberation measurement for this purpose was an experimental *tour de force* carried out almost a decade before vacuum tubes and microphones had made it feasible to measure sound pressures directly. Sabine's paper on "Whispering Galleries" was still in manuscript when he died and it had its only publication in this collection. He dismisses most of these acoustical curiosities as accidents, but what he wrote here about six of the famous ones is still definitive.

Sabine's academic career took a new turn in 1906 when Harvard's Lawrence Scientific School evolved into a Graduate School of Applied Science and Sabine became its Dean. He discharged these duties with distinction for nine years until the post itself dissolved in the short-lived Harvard–Massachusetts Institute of Technology merger of 1915–1917. Relatively few of these *Papers* appeared during the interim between 1900 and 1912, in part because Sabine had an uncommon reluctance to publish the results of an experiment until he was sure it had been done so well that nobody would ever need to repeat it. Two incidents occurred in 1911, however, that may have provided some motivation for the preparation of his summary papers of the 1912–1915 period: Sabine became involved in a patent case, and he suffered a mild cerebral hemorrhage from which he recovered only to a precarious state of good health. Jacob Mazer, a Pittsburgh architectural consultant to whom Sabine had previously explained his principles of reverberation control, sought in his own name a patent covering these principles and the materials Sabine had used for sound absorption. Only prompt intercession by the President of the United States prevented the issuance of this patent that would have denied to Sabine and others the free use of the discoveries he thought he had dedicated to the public. Somewhat later, and perhaps on account of this bitter experience, Sabine did allow his name to be attached to two patent applications, one for a ceramic tile and one for a cloth-covered felt tile (Akoustolith).

World War I provided an opportunity for Sabine to contribute to a cause about which he felt deeply. Although he had predicted the imminence of a German war as early as 1909, the summer of 1914 found him and his two daughters in Germany again. They were barely able to escape through Holland to England on 3 August, on the very eve of the German invasion of Belgium. After another academic year at home, during which he wrote his final summary paper on "Architectural Acoustics," Sabine returned to Europe and to the preoccupation with World War I that consumed the remainder of his life.

The field of architectural acoustics has grown in breadth and in depth since the date of Sabine's last paper. New and improved acoustical materials have been developed and acoustic-impedance concepts have opened the way to better understanding of the physics of sound absorption. Normal mode analysis has provided a new conceptual framework in which to consider the behavior of sound in rooms, and there has grown up a new appreciation of the effects of sound diffusion and of the beneficial reinforcement provided by early reflections. The growth and decay of sound in empty rectangular rooms can now be analyzed in precise detail, and there is some hope that modern computers will make it possible to account in similar detail for the acoustical performance of real rooms with irregular boundaries and interior furnishings. Nevertheless, in spite of these substantial increments of understanding, Sabine's hyperbola is still the hard core of the science of sound control in listening rooms. This simple relation, in the form of an equation for reverberation time, continues to serve as the basic analytical tool of the consultant who seeks to quiet a noisy office or to make a school room or church acoustically tolerable.

International standardization has honored Sabine by naming the area unit of sound absorption after him. Bronze plaques memorialize him in Symphony Hall and in the lecture room of the Jefferson Laboratory, but later generations may remember him even longer and more fondly for his hyperbola and for these *Collected Papers*.

FREDERICK V. HUNT

HARVARD UNIVERSITY
January, 1964

PREFACE

THIS volume aims to contain all the important contributions to the subject of acoustics from the pen of the late Professor W. C. Sabine. The greater part of these papers appeared in a number of different architectural journals and were therefore addressed to a changing audience, little acquainted with physical science, and to whose members the subject was altogether novel. Under these circumstances a certain amount of repetition was not only unavoidable, but desirable. Little attempt has been made to reduce this repetition but in one case an omission seemed wise. The material contained in the author's earliest papers on acoustics, which appeared in the *Proceedings of the American Institute of Architects* in 1898, is repeated almost completely in the paper which forms the first chapter of this volume; it has, therefore, been omitted from this collection with the exception of a few extracts which have been inserted as footnotes in the first chapter.

No apology is made for the preservation of the paper from the *Proceedings of the Franklin Institute*, for, though much of the material therein is to be found in the earlier chapters of this volume, the article is valuable as a summary, and as such it is recommended to the reader who desires to obtain a general view of the subject.

In addition to the papers already in print at the time of the author's death the only available material consisted of the manuscripts of two articles, one on Echoes, the other on Whispering Galleries, and the full notes on four of the lectures on acoustics delivered at the Sorbonne in the spring of 1917. Of this material, the first paper was discarded as being too fragmentary; the second, after some slight omissions and corrections in the text made necessary by the loss of a few of the illustrations, forms Chapter 11 of this volume; an abstract of so much of the substance of the lecture notes as had not already appeared in print has been made, of which part is to be found in the form of an Appendix and part is contained in some of the following paragraphs.

The reader may often be puzzled by reference to works about to be published but of which no trace is to be found in this volume. It is a melancholy fact that these papers were either never written or else were destroyed by their author; no trace of them can be found. The extent of the labors of which no adequate record remains may best be judged from the following extracts taken from the notes on the Paris lectures just mentioned.

" On the one hand we have the problem (Reverberation) which we have been discussing up to the present moment, and on the other the whole question of the transmission of sound from one room to another, through the walls, the doors, the ceiling and the floors; and the telephonic transmission, if I may so call it, through the length of the structure. It is five years ago since this second problem was first attacked and though the research is certainly not complete, some ground has been covered. A quantitatively exact method has been established and the transmission of sound through about twenty different kinds of partitions has been determined.

" For example: Transmission of sound through four kinds of doors has been studied; two of oak, two of pine, one of each kind was paneled and was relatively thin and light; one of each kind was very heavy, nearly four centimetres thick; through four kinds of windows, one of plate glass, one with common panes, one double with an air space of two centimetres between, one with small panes set in lead such as one sees in churches; through brick walls with plaster on both sides; through walls of tile similarly plastered; through walls of a character not common in France and which we call gypsum block; through plaster on lath; through about ten different kinds of sound insulators, patented, and sold in quantities representing hundreds of thousands of dollars each year, yet practically without value, since one can easily converse through six thicknesses of these substances and talk in a low tone through three, while a single thickness is that ordinarily employed. The behavior of an air space has been studied, the effect of the thickness of this air space, and the result of filling the space with sand, saw-dust and asbestos. In spite of all this, the research is far from complete and many other forms of construction must be investigated before it will be possible to publish the results; these determinations must be made with the greatest exactness as very important interests are involved. . . .

"The research is particularly laborious because resonance has a special importance in a great number of forms of construction. It is a much greater factor in transmission than in absorption.

"I shall not enlarge on this subject here for two reasons: first, I believe that it is not of special interest, at least, in its present state, and second, because it is not proper to present a formal discussion of this subject while the research is still unfinished."

The last paragraph is characteristic. The severity of the criticism which Professor Sabine always applied to his own productions increased with time, and it is to this extreme self-criticism and repression that we must ascribe the loss of much invaluable scientific material.

Thanks are due to The American Institute of Architects and to the editors of *The American Architect*, *The Brickbuilder*, *The Engineering Record*, and *The Journal of the Franklin Institute*, for permission to reprint the articles which originally appeared in their respective Journals.

The Editor is also greatly obliged to Dr. Paul Sabine and Mr. Clifford M. Swan for a great deal of valuable material, and to Mr. Frank Chouteau Brown for his assistance in seeing the book through the press. He is particularly indebted to his colleague Professor F. A. Saunders for his invaluable aid in all matters touching the correct presentation of the material of this volume.

Theodore Lyman

JEFFERSON PHYSICAL LABORATORY
Harvard University
June, 1921

CONTENTS

COLLECTED PAPERS
ON ACOUSTICS

1

REVERBERATION[1]

INTRODUCTION

THE following investigation was not undertaken at first by choice, but devolved on the writer in 1895 through instructions from the Corporation of Harvard University to propose changes for remedying the acoustical difficulties in the lecture-room of the Fogg Art Museum, a building that had just been completed. About two years were spent in experimenting on this room, and permanent changes were then made. Almost immediately afterward it became certain that a new Boston Music Hall would be erected, and the questions arising in the consideration of its plans forced a not unwelcome continuance of the general investigation.

No one can appreciate the condition of architectural acoustics — the science of sound as applied to buildings — who has not with a pressing case in hand sought through the scattered literature for some safe guidance. Responsibility in a large and irretrievable expenditure of money compels a careful consideration, and emphasizes the meagerness and inconsistency of the current suggestions. Thus the most definite and often repeated statements are such as the following, that the dimensions of a room should be in the ratio 2 : 3 : 5, or according to some writers, 1 : 1 : 2, and others, 2 : 3 : 4; it is probable that the basis of these suggestions is the ratios of the harmonic intervals in music, but the connection is untraced and remote. Moreover, such advice is rather difficult to apply; should one measure the length to the back or to the front of the galleries, to the back or the front of the stage recess? Few rooms have a flat roof, where should the height be measured? One writer, who had seen the Mormon Temple, recommended that all auditoriums be elliptical. Sanders Theatre is by far the best auditorium in Cambridge and is semicircular in general shape, but with a recess that makes it almost anything; and, on the other hand, the lecture-room in the Fogg Art

[1] The American Architect and The Engineering Record, 1900.

Museum is also semicircular, indeed was modeled after Sanders Theatre, and it was the worst. But Sanders Theatre is in wood and the Fogg lecture-room is plaster on tile; one seizes on this only to be immediately reminded that Sayles Hall in Providence is largely lined with wood and is bad. Curiously enough, each suggestion is advanced as if it alone were sufficient. As examples of remedies, may be cited the placing of vases about the room for the sake of resonance, wrongly supposed to have been the object of the vases in Greek theatres, and the stretching of wires, even now a frequent though useless device.

The problem is necessarily complex, and each room presents many conditions, each of which contributes to the result in a greater or less degree according to circumstances. To take justly into account these varied conditions, the solution of the problem should be quantitative, not merely qualitative; and to reach its highest usefulness it should be such that its application can precede, not follow, the construction of the building.

In order that hearing may be good in any auditorium, it is necessary that the sound should be sufficiently loud; that the simultaneous components of a complex sound should maintain their proper relative intensities; and that the successive sounds in rapidly moving articulation, either of speech or music, should be clear and distinct, free from each other and from extraneous noises. These three are the necessary, as they are the entirely sufficient, conditions for good hearing. The architectural problem is, correspondingly, threefold, and in this introductory paper an attempt will be made to sketch and define briefly the subject on this basis of classification. Within the three fields thus defined is comprised without exception the whole of architectural acoustics.

1. *Loudness*. — Starting with the simplest conceivable auditorium — a level and open plain, with the ground bare and hard, a single person for an audience — it is clear that the sound spreads in a hemispherical wave diminishing in intensity as it increases in size, proportionally. If, instead of being bare, the ground is occupied by a large audience, the sound diminishes in intensity even more rapidly, being now absorbed. The upper part of the sound-wave escapes unaffected, but the lower edge — the only part that is of service to an

audience on a plain — is rapidly lost. The first and most obvious improvement is to raise the speaker above the level of the audience; the second is to raise the seats at the rear; and the third is to place a wall behind the speaker. The result is most attractively illustrated in the Greek theatre. These changes being made, still all the sound rising at any considerable angle is lost through the opening above, and only part of the speaker's efforts serve the audience. When to this auditorium a roof is added the average intensity of sound throughout the room is greatly increased, especially that of sustained tones; and the intensity of sound at the front and the rear is more nearly equalized. If, in addition, galleries be constructed in order to elevate the distant part of the audience and bring it nearer to the front, we have the general form of the modern auditorium. The problem of calculating the loudness at different parts of such an auditorium is, obviously, complex, but it is perfectly determinate, and as soon as the reflecting and absorbing power of the audience and of the various wall-surfaces are known it can be solved approximately. Under this head will be considered the effect of sounding-boards, the relative merits of different materials used as reflectors, the refraction of sound, and the influence of the variable temperature of the air through the heating and ventilating of the room, and similar subjects.

2. *Distortion of Complex Sounds: Interference and Resonance.* — In discussing the subject of loudness the direct and reflected sounds have been spoken of as if always reënforcing each other when they come together. A moment's consideration of the nature of sound will show that, as a matter of fact, it is entirely possible for them to oppose each other. The sounding body in its forward motion sends off a wave of condensation, which is immediately followed through the air by a wave of rarefaction produced by the vibrating body as it moves back. These two waves of opposite character taken together constitute a sound-wave. The source continuing to vibrate, these waves follow each other in a train. Bearing in mind this alternating character of sound, it is evident that should the sound traveling by different paths — by reflection from different walls — come together again, the paths being equal in length, condensation will arrive at the same time as condensation, and reënforce it, and rare-

faction will, similarly, reënforce rarefaction. But should one path be a little shorter than the other, rarefaction by one and condensation by the other may arrive at the same time, and at this point there will be comparative silence. The whole room may be mapped out into regions in which the sound is loud and regions in which it is feeble. When there are many reflecting surfaces the interference is much more complex. When the note changes in pitch the interference system is entirely altered in character. A single incident will serve to illustrate this point. There is a room in the Jefferson Physical Laboratory, known as the constant-temperature room, that has been of the utmost service throughout these experiments. It is in the center of one wing of the building, is entirely under ground, even below the level of the basement of the building, has separate foundations and double walls, each wall being very thick and of brick in cement. It was originally designed for investigations in heat requiring constant temperature, and its peculiar location and construction were for this purpose. As it was not so in use, however, it was turned over to these experiments in sound, and a room more suitable could not be designed. From its location and construction it is extremely quiet. Without windows, its walls, floor, and ceiling — all of solid masonry — are smooth and unbroken. The single door to the room is plain and flush with the wall. The dimensions of the room are, on the floor, 4.27 × 6.10 meters; its height at the walls is 2.54 meters, but the ceiling is slightly arched, giving a height at the center of 3.17 meters. This room is here described at length because it will be frequently referred to, particularly in this matter of interference of sound. While working in this room with a treble c gemshorn organ pipe blown by a steady wind-pressure, it was observed that the pitch of the pipe apparently changed an octave when the observer straightened up in his chair from a position in which he was leaning forward. The explanation is this: The organ pipe did not give a single pure note, but gave a fundamental treble c accompanied by several overtones, of which the strongest was in this case the octave above. Each note in the whole complex sound had its own interference system, which, as long as the sound remained constant, remained fixed in position. It so happened that at these two points the region of silence for one note coincided with the region of reënforcement in the other, and *vice versa*. Thus the observer in one position heard the fundamental note, and in the other, the first overtone. The change was exceedingly striking, and as the notes remained constant, the experiment could be tried again and again. With a little search it was possible to find other points in the room at which the same phenomenon appeared, but generally in less perfection. The distortion of the relative intensities of the components of a chord that may thus be produced is evident. Practically almost every sound of the voice in speech and song, and of instrumental music, even single-part music so-called, is more or less complex, and, therefore, subject to this distortion. It will be necessary, later, to show under what circumstances this phenomenon is a formidable danger, and how it may be guarded against, and under what circumstances it is negligible. It is evident from the above occurrence that it may be a most serious matter, for in this room two persons side by side can talk together with but little comfort, most of the difficulty being caused by the interference of sound.

There is another phenomenon, in its occurrence allied to interference, but in nature distinct — the phenomenon of resonance. Both, however, occasion the same evil — the distortion of that nice adjustment of the relative intensities of the components of the complex sounds that constitute speech and music. The phenomenon of interference just discussed merely alters the distribution of sound in the room, causing the intensity of any one pure sustained note to be above or below the average intensity at near points. Resonance, on the other hand, alters the total amount of sound in the whole room and always increases it. This phenomenon is noticeable at times in using the voice in a small room, or even in particular locations in a large room. Perhaps its occurrence is most easily observed in setting up a large church organ, where the pipes must be readjusted for the particular space in which the organ is to stand, no matter with how much care the organ may have been assembled and adjusted before leaving the factory. The general phenomenon of resonance is of very wide occurrence, not merely in acoustics but in more gross mechanics as well, as the vibration of a bridge to a properly timed tread, or the excessive rolling of a boat

in certain seas. The principle is the same in all cases. The following conception is an easy one to grasp, and is closely analogous to
acoustical resonance: If the palm of the hand be placed on the
center of the surface of water in a large basin or tank and quickly
depressed and raised once it will cause a wave to spread, which,
reflected at the edge of the water, will return, in part at least, to
the hand. If, just as the wave reaches the hand, the hand repeats
its motion with the same force, it will reënforce the wave traveling
over the water. Thus reënforced, the wave goes out stronger than
before and returns again. By continued repetition of the motion
of the hand so timed as to reënforce the wave as it returns, the wave
gets to be very strong. Instead of restraining the hand each time
until the wave traveling to and fro returns to it, one may so time
the motion of the hand as to have several equal waves following
each other over the water, and the hand each time reënforcing the
wave that is passing. This, obviously, can be done by dividing the
interval of time between the successive motions of the hand by any
whole number whatever, and moving the hand with the frequency
thus defined. The result will be a strong reënforcement of the waves.
If, however, the motions of the hand be not so timed, it is obvious
that the reënforcement will not be perfect, and, in fact, it is possible
to so time it as exactly to oppose the returning waves. The application of this reasoning to the phenomenon of sound, where the air
takes the place of the water and the sounding body that of the hand,
needs little additional explanation. Some notes of a complex sound
are reënforced, some are not, and thus the quality is altered. This
phenomenon enters in two forms in the architectural problem: there
may be either resonance of the air in the room or resonance of the
walls, and the two cases must receive separate discussion; their
effects are totally different.

The word "resonance" has been used loosely as synonymous
with "reverberation," and even with "echo," and is so given in
some of the more voluminous but less exact popular dictionaries.
In scientific literature the term has received a very definite and
precise application to the phenomenon, wherever it may occur, of
the growth of a vibratory motion of an elastic body under periodic
forces timed to its natural rates of vibration. A word having this

significance is necessary; and it is very desirable that the term should not, even popularly, by meaning many things, cease to mean anything exactly.

3. *Confusion: Reverberation, Echo and Extraneous Sounds.* — Sound, being energy, once produced in a confined space, will continue until it is either transmitted by the boundary walls, or is transformed into some other kind of energy, generally heat. This process of decay is called absorption. Thus, in the lecture-room of Harvard University, in which, and in behalf of which, this investigation was begun, the rate of absorption was so small that a word spoken in an ordinary tone of voice was audible for five and a half seconds afterwards. During this time even a very deliberate speaker would have uttered the twelve or fifteen succeeding syllables. Thus the successive enunciations blended into a loud sound, through which and above which it was necessary to hear and distinguish the orderly progression of the speech. Across the room this could not be done; even near the speaker it could be done only with an effort wearisome in the extreme if long maintained. With an audience filling the room the conditions were not so bad, but still not tolerable. This may be regarded, if one so chooses, as a process of multiple reflection from walls, from ceiling and from floor, first from one and then another, losing a little at each reflection until ultimately inaudible. This phenomenon will be called reverberation, including as a special case the echo. It must be observed, however, that, in general, reverberation results in a mass of sound filling the whole room and incapable of analysis into its distinct reflections. It is thus more difficult to recognize and impossible to locate. The term echo will be reserved for that particular case in which a short, sharp sound is distinctly repeated by reflection, either once from a single surface, or several times from two or more surfaces. In the general case of reverberation we are only concerned with the rate of decay of the sound. In the special case of the echo we are concerned not merely with its intensity, but with the interval of time elapsing between the initial sound and the moment it reaches the observer. In the room mentioned as the occasion of this investigation, no discrete echo was distinctly perceptible, and the case will serve excellently as an illustration of the more general

type of reverberation. After preliminary gropings,[1] first in the literature and then with several optical devices for measuring the intensity of sound, both were abandoned, the latter for reasons that will be explained later. Instead, the rate of decay was measured by measuring what was inversely proportional to it — the duration of audibility of the reverberation, or, as it will be called here, the duration of audibility of the residual sound. These experiments may be explained to advantage even in this introductory paper, for they will give more clearly than would abstract discussion an idea of the nature of reverberation. Broadly considered, there are two, and only two, variables in a room — shape including size, and materials including furnishings. In designing an auditorium an architect can give consideration to both; in repair work for bad acoustical conditions it is generally impracticable to change the shape, and only variations in materials and furnishings are allowable. This was, therefore, the line of work in this case. It was evident that, other things being equal, the rate at which the reverberation would disappear was proportional to the rate at which the sound was absorbed. The first work, therefore, was to determine the relative absorbing power of various substances. With an organ pipe as a constant source of sound, and a suitable chronograph for recording, the duration of audibility of a sound after the source had ceased in this room when empty was found to be 5.62 seconds. All the cushions from the seats in Sanders Theatre were then brought over and stored in the lobby. On bringing into the lecture-room a number of cushions having a total length of 8.2 meters, the duration of audibility fell to 5.33 seconds. On bringing in 17 meters the sound in the room after the organ pipe ceased was audible for but 4.94

[1] The first method for determining the rate of decay of the sound, and therefore the amount of absorption, was by means of a sensitive manometric gas flame measured by a micrometer telescope. Later, photographing the flame was tried; but both methods were abandoned, for they both showed, what the unaided ear could perceive, that the sound as observed at any point in the room died away in a fluctuating manner, passing through maxima and minima. Moreover, they showed what the unaided ear had not detected, but immediately afterward did recognize, that the sound was often more intense immediately after the source ceased than before. All this was interesting, but it rendered impossible any accurate interpretation of the results obtained by these or similar methods. It was then found that the ear itself aided by a suitable electrical chronograph for recording the duration or audibility of the residual sound gave a surprisingly sensitive and accurate method of measurement. Proc. American Institute of Architects, p. 35, 1898.

seconds. Evidently, the cushions were strong absorbents and rapidly improving the room, at least to the extent of diminishing the reverberation. The result was interesting and the process was continued. Little by little the cushions were brought into the room, and each time the duration of audibility was measured. When all the seats (436 in number) were covered, the sound was audible for 2.03 seconds. Then the aisles were covered, and then the platform. Still there were more cushions — almost half as many more. These were brought into the room, a few at a time, as before, and draped on a scaffolding that had been erected around the room, the duration of the sound being recorded each time. Finally, when all the cushions from a theatre seating nearly fifteen hundred persons were placed in the room — covering the seats, the aisles, the platform, the rear wall to the ceiling — the duration of audibility of the residual sound was 1.14 seconds. This experiment, requiring, of course, several nights' work, having been completed, all the cushions were removed and the room was in readiness for the test of other absorbents. It was evident that a standard of comparison had been established. Curtains of chenille, 1.1 meters wide and 17 meters in total length, were draped in the room. The duration of audibility was then 4.51 seconds. Turning to the data that had just been collected it appeared that this amount of chenille was equivalent to 30 meters of Sanders Theatre cushions. Oriental rugs, Herez, Demirjik, and Hindoostanee, were tested in a similar manner; as were also cretonne cloth, canvas, and hair felt. Similar experiments, but in a smaller room, determined the absorbing power of a man and of a woman, always by determining the number of running meters of Sanders Theatre cushions that would produce the same effect. This process of comparing two absorbents by actually substituting one for the other is laborious, and it is given here only to show the first steps in the development of a method that will be expanded in the following papers.

In this lecture-room felt was finally placed permanently on particular walls, and the room was rendered not excellent, but entirely serviceable, and it has been used for the past three years without serious complaint. It is not intended to discuss this particular case in the introductory paper, because such discussion would be premature and logically incomplete. It is mentioned here merely to illustrate concretely the subject of reverberation, and its dependence on absorption. It would be a mistake to suppose that an absorbent is always desirable, or even when desirable that its position is a matter of no consequence.[1]

While the logical order of considering the conditions contributing to or interfering with distinct hearing would be that employed above, it so happens that exactly the reverse order is preferable from an experimental standpoint. By taking up the subject of reverberation first it is possible to determine the coefficients of absorption and reflection of various kinds of wall surface, of furniture and draperies, and of an audience. The investigation of reverberation is now, after five years of experimental work, completed, and an account will be rendered in the following papers. Some data have also been secured on the other topics and will be published as soon as rounded into definite form.

This paper may be regarded as introductory to the general subject of architectural acoustics, and immediately introductory to a series of articles dealing with the subject of reverberation, in which the general line of procedure will be, briefly, as follows: The absorbing power of wall-surfaces will be determined, and the law according to which the reverberation of a room depends on its volume will be demonstrated. The absolute rate of decay of the residual sound in a number of rooms, and in the same room under different conditions, will then be determined. In the fifth paper a more exact analysis

[1] There is no simple treatment that can cure all cases. There may be inadequate absorption and prolonged residual sound; in this case absorbing material should be added in the proper places. On the other hand, there may be excessive absorption by the nearer parts of the hall and by the nearer audience and the sound may not penetrate to the greater distances. Obviously the treatment should not be the same. There is such a room belonging to the University, known locally as Sever 35. It is low and long. Across its ceiling are now stretched hundreds of wires and many yards of cloth. The former has the merit of being harmless, the latter is like bleeding a patient suffering from a chill. In general, should the sound seem smothered or too faint, it is because the sound is either imperfectly distributed to the audience, or is lost in waste places. The first may occur in a very low and long room, the second in one with a very high ceiling. The first can be remedied only slightly at best, the latter can be improved by the use of reflectors behind and above the speaker. On the other hand, should the sound be loud but confused, due to a perceptible prolongation, the difficulty arises from there being reflecting surfaces either too far distant or improperly inclined. Proc. American Institute of Architects, p. 39, 1898.

will be given, and it will be shown that, by very different lines of attack, starting from different data, the same numerical results are secured. Tables will be given of the absorbing power of various wall-surfaces, of furniture, of an audience, and of all the materials ordinarily found in any quantity in an auditorium. Finally, in illustration of the calculation of reverberation in advance of construction, will be cited the new Boston Music Hall, the most interesting case that has arisen.

ABSORBING POWER OF WALL–SURFACES

In the introductory article the problem was divided into considerations of loudness, of distortion, and of confusion of sounds. Confusion may arise from extraneous disturbing sounds — street noises and the noise of ventilating fans — or from the prolongation of the otherwise discrete sounds of music or the voice into the succeeding sounds. The latter phenomenon, known as reverberation, results in what may be called, with accuracy and suggestiveness, residual sound. The duration of this residual sound was shown to depend on the amount of absorbing material inside the room, and also, of course, on the absorbing and transmitting power of the walls; and a method was outlined for determining the absorbing power of the former in terms of the absorbing power of some material chosen as a standard and used in a preliminary calibration. A moment's consideration demonstrates that this method, which is of the general type known as a "substitution method," while effective in the determination of the absorbing power of furniture and corrective material, and, in general, of anything that can be brought into or removed from a room, is insufficient for determinating the absorbing power of wall-surfaces. This, the absorbing power of wall-surfaces, is the subject of the present paper; and as the method of determination is an extension of the above work, and finds its justification in the striking consistency of the results of the observations, a more elaborate description of the experimental method is desirable. A proof of the accuracy of every step taken is especially necessary in a subject concerning which theory has been so largely uncontrolled speculation.

Early in the investigation it was found that measurements of the length of time during which a sound was audible after the source had ceased gave promising results whose larger inconsistencies could be traced directly to the distraction of outside noises. On repeating the work during the most quiet part of the night, between half-past twelve and five, and using refined recording apparatus, the minor irregularities, due to relaxed attention or other personal variations, were surprisingly small. To secure accuracy, however, it was necessary to suspend work on the approach of a street car within two blocks, or on the passing of a train a mile distant. In Cambridge these interruptions were not serious; in Boston and in New York it was necessary to snatch observations in very brief intervals of quiet. In every case a single determination of the duration of the residual sound was based on the average of a large number of observations.

An organ pipe, of the gemshorn stop, an octave above middle *c* (512 vibration frequency) was used as the source of sound in some preliminary experiments, and has been retained in subsequent work in the absence of any good reason for changing. The wind supply from a double tank, water-sealed and noiseless, was turned on and off the organ pipe by an electro-pneumatic valve, designed by Mr. George S. Hutchings, and similar to that used in his large church organs. The electric current controlling the valve also controlled the chronograph, and was made and broken by a key in the hands of the observer from any part of the room. The chronograph employed in the later experiments, after the more usual patterns had been tried and discarded, was of special design, and answered well the requirements of the work — perfect noiselessness, portability, and capacity to measure intervals of time from a half second to ten seconds with considerable accuracy. It is shown in the adjacent diagram. The current whose cessation stopped the sounding of the organ pipe also gave the initial record on the chronograph, and the only duty of the observer was to make the record when the sound ceased to be audible.

While the supreme test of the investigation lies in the consistency and simplicity of the whole solution as outlined later, three preliminary criteria are found in (1) the agreement of the observations

obtained at one sitting, (2) the agreement of the results obtained
on different nights and after the lapse of months, or even years, by
the same observer under similar conditions, and (3) the agreement
of independent determinations by different observers. The first
can best be discussed, of course, by the recognized physical methods
for examining the accuracy of an extended series of observations;

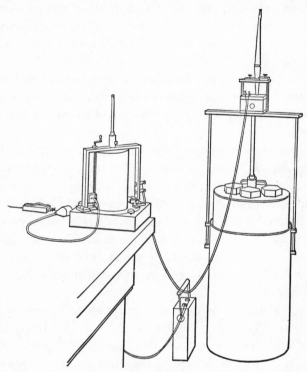

Fig. 1.   Chronograph, battery, and air reservoir, the latter surmounted
by the electro-pneumatic valve and organ pipe.

and the result of such examination is as follows: Each determination
being the mean of about twenty observations under conditions such
that the audible duration of the residual sound was 4 seconds, the
average deviation of the single observations from the mean was .11
seconds, and the maximum deviation was .31.   The computed
"probable error" of a single determination was about .02 seconds;
as a matter of fact, the average deviation of ten determinations
from the mean of the ten was .03 seconds, and the maximum devi-

ation was .05. The reason for this accuracy will be discussed in a subsequent paper. The probable error of the mean, thus calculated from the deviations of the single observations, covers only those variable errors as likely to increase as to decrease the final result. Fixed instrumental errors, and the constant errors commonly referred to by the term "personal factors" are not in this way exposed. They were, however, repeatedly tested for by comparison with a clock beating seconds, and were very satisfactorily shown not to amount to more than .02 seconds in their cumulative effect. Three types of chronographs, and three kinds of valves between the organ pipe and the wind chest were used in the gradual development of the experiment, and all gave for the same room very nearly the same final results. The later instruments were, of course, better and more accurate.

The second criterion mentioned above is abundantly satisfied by the experiments. Observations taken every second or third night for two months in the lecture-room of the Fogg Art Museum gave practically the same results, varying from 5.45 to 5.62 with a mean value of 5.57 seconds, a result, moreover, that was again obtained after the lapse of one and then of three years. Equally satisfactory agreement was obtained at the beginning and at the end of three years in Sanders Theatre, and in the constant-temperature room of the Physical Laboratory.

Two gentlemen, who were already somewhat skilled in physical observation, Mr. Gifford LeClear and Mr. E. D. Densmore, gave the necessary time to test the third point. After several nights' practice their results differed but slightly, being .08 seconds and .10 seconds longer than those obtained by the writer, the total duration of the sound being 4 seconds. This agreement, showing that the results are probably very nearly those that would be obtained by any auditor of normal hearing, gives to them additional interest. It should be stated, however, that the final development of the subject will adapt it with perfect generality to either normal or abnormal acuteness of hearing.

Almost the first step in the investigation was to establish the following three fundamentally important facts. Later work has proved these fundamental facts far more accurately, but the original experiments are here given as being those upon which the conclusions were based.

The duration of audibility of the residual sound is nearly the same in all parts of an auditorium. — Early in the investigation an experiment to test this point was made in Steinert Hall, in Boston. The source of sound remaining on the platform at the point marked

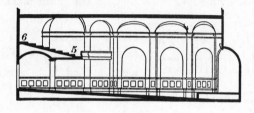

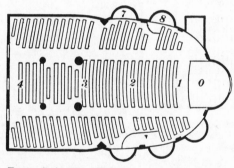

Fig. 2.   Steinert Hall, Boston: position of air reservoir and organ pipe at *O*; positions of observer 1-8.

O in the diagram, observations were made in succession at the points marked 1 to 8, with the results shown in the table:

Station	Duration	Station	Duration
1	2.12	5	2.23
2	2.17	6	2.27
3	2.23	7	2.20
4	2.20	8	2.26

On first inspection these results seem to indicate that the duration of audibility is very slightly greater at a distance from the source, and it would be easy to explain this on the theory that at a distance the ear is less exhausted by the rather loud noise while the pipe is sounding; but, as a matter of fact, this is not the case, and the variations there shown are within the limits of accuracy of the apparatus employed and the skill attained thus early in the investigation. Numerous later experiments, more accurate, but not especially directed to this point, have verified the above general statement quite conclusively.

The duration of audibility is nearly independent of the position of the source. — The observer remaining at the point marked *O* in the diagram of the large lecture-room of the Jefferson Physical Laboratory, the organ pipe and wind chest were moved from station to station, as indicated by the numbers 1 to 6, with the results shown in the table:

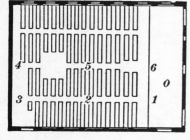

FIG. 3. Lecture-room, Jefferson Physical Laboratory: position of observer at *O*; positions of air reservoir and organ pipe 1-6.

Station	Duration
1	3.90
2	4.00
3	3.90
4	3.98
5	3.95
6	3.96

The efficiency of an absorbent in reducing the duration of the residual sound is, under ordinary circumstances, nearly independent of its position. — Fifty meters of cretonne cloth draped on a scaffolding under the rather low ceiling at the back of the lecture-room of the Fogg Museum, as shown in the next diagram, reduced the audible duration of the residual sound by very nearly the same amount, regardless of the section in which it hung, as shown in the following table, the initial duration being 5.57 seconds:

Section	Duration
1	4.88
2	4.83
3	4.92
4	4.85

In some later experiments five and a half times as much cretonne draped on the scaffolding reduced the audible duration of the residual sound to 3.25 seconds; and when hung fully exposed in the high dome-like ceiling, gave 3.29 seconds, confirming the above statement.

These facts, simple when proved, were by no means self-evident so long as the problem was one of reverberation, that is, of succes_sive reflection of sound from wall to wall. They indicated that, at least with reference to auditoriums of not too great dimensions, another point of view would be more suggestive, that of regarding the whole as an energy problem in which the source is at the organ pipe and the decay at the walls and at the contained absorbing material. The above results, then, all point to the evident, but perhaps not appreciated, fact that the dispersion of sound between all parts of a hall is very rapid in comparison with the total time required for its complete absorption, and that in a very short time after the source has ceased the intensity of the residual sound, except for the phenomenon of interference to be considered later, is very nearly the same everywhere in the room.

This much being determined, the investigation was continued in the following manner: Cushions from Sanders Theatre were transferred to the lobby of the lecture-room of the Fogg

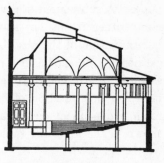

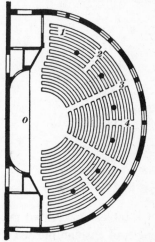

Fig. 4. Lecture-room, Fogg Art Museum: position of observer at O; positions of absorbent at 1-4, and in the dome.

Museum; a very few were brought into the room and spread along the front row of seats; the duration of audibility of the residual sound, diminished by the introduction of this additional absorbent, was determined, and the total length of cushion was measured. The next row of seats was then covered in the same manner and the two observations made — length of cushion and duration of residual sound. This was repeated till cushions covered all the seats. This work was at first undertaken solely with the intention of determining the relative merits of different absorbing materials that might be placed in the room as a corrective for excessive residual sound, and the account of this application is given in the introductory paper. A subsequent study of these and similar results obtained in many other rooms has shown their applicability to the accurate determination of the absorbing power of wall-surfaces. This application may be shown in a purely analytical manner, but the exposition is greatly helped by a graphical representation. The manner in which the duration of the residual sound in the Fogg lecture-room is dependent on the amount of absorbing material present is shown in the following table:

Length of Cushion in Meters	Duration of Residual Sound in Seconds
0	5.61
8	5.33
17	4.94
28	4.56
44	4.21
63	3.94
83	3.49
104	3.33
128	3.00
145	2.85
162	2.64
189	2.36
213	2.33
242	2.22

This table, represented graphically in the conventional manner — length of cushion plotted horizontally and duration of sound vertically — gives points through which the curve may be drawn in the accompanying diagram. To discover the law from this curve we represent the lengths of cushion by x, and the corresponding durations of sound, the vertical distances to the curve, by t. If we now seek the formula connecting x and t that most nearly expresses the relationship represented by the above curve, we find it to be $(a + x)t = k$, which is the familiar formula of a rectangular hyperbola with its origin displaced along the axis of x, one of its asymptotes, by an amount a. To make this formula most closely fit our curve we must, in this case, give to the constant, *a*, the numerical value, 146, and to *k* the value, 813. The accuracy with which the formula represents the curve may be seen by comparing the durations calculated by the formula with those determined from the curve; they nowhere differ by more than .04 of a second, and have, on an average, a difference of only .02 of a second. This is entirely satisfactory, for the calculated points fall off from the curve by scarcely the breadth of the pen point with which it was drawn.

The determination of the absorbing power of the wall-surface depends on the interpretation of the constant, *a*. In the formula,

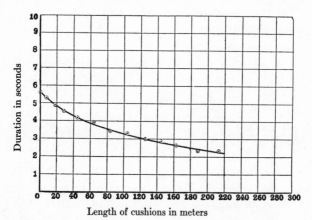

FIG. 5. Curve showing the relation of the duration of the residual
sound to the added absorbing material.

the position of *a*, indicating that *x* is to be added to it, suggests that *x* and *a* are of a like nature, and that *a* is a measure of the absorbing power of the bare room; in order to determine the curve this was increased by the introduction of the cushions. This is even better shown by the diagram in which the portion of the curve experimentally determined is fitted into the curve as a whole, and *a* and *x* are indicated. Thus, the absorbing power of the room — the walls, partly plaster on stone, partly plaster on wire lath, the windows, the skylight, the floor — was equivalent to 146 running meters of Sanders Theatre cushions.

The last statement shows the necessity for two subsidiary investigations. The first, to express the results in some more permanent, more universally available, and, if possible, more absolute unit than the cushions; the other, to apportion the total absorbing power among the various components of the structure.

The transformation of results from one system of units to another necessitates a careful study of both systems. Some early experiments in which the cushions were placed with one edge pushed against the backs of the settees gave results whose anomalous character suggested that, perhaps, their absorbing power depended not merely on the amount present but also on the area of the surface exposed. It was then recalled that about two years before, at the beginning of an evening's work, the first lot of cushions

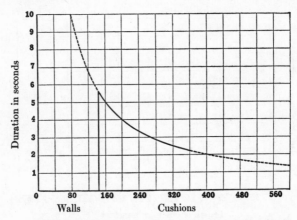

FIG. 6. Curve 5 plotted as part of its corresponding rectangular hyperbola. The solid part was determined experimentally; the displacement of this to the right measures the absorbing power of the walls of the room.

brought into the room were placed on the floor, side by side, with edges touching, but that after a few observations had been taken the cushions were scattered about the room, and the work was repeated. This was done not at all to uncover the edges, but in the primitive uncertainty as to whether near cushions would draw from each other's supply of sound, as it were, and thus diminish each other's efficiency. No further thought was then given to these discarded observations until recalled by the above-mentioned discrepancy. They were sought out from the notes of that period, and it was found that, as suspected, the absorbing power of the cushions when touching edges was less than when separated. Eight cushions had been used, and, therefore, fourteen edges had been touching. A record was found of the length and the breadth of the cushions used, and, assuming that the absorbing power was proportional to the area exposed, it was possible to calculate their thickness by comparing the audible duration of the residual sound in the two sets of observations; it was thus calculated to be 7.4 centimeters. On stacking up the same cushions and measuring their total thickness, the average thickness was found to be 7.2 centimeters, in very close agreement with the thickness estimated from their absorption of sound. Therefore, the measurements of the cushions should be, not in running meters of cushion, but in square meters of exposed surface.

For the purposes of the present investigation, it is wholly unnecessary to distinguish between the transformation of the energy of the sound into heat and its transmission into outside space. Both shall be called absorption. The former is the special accomplishment of cushions, the latter of open windows. It is obvious, however, that if both cushions and windows are to be classed as absorbents, the open window, because the more universally accessible and the more permanent, is the better unit. The cushions, on the other hand, are by far the more convenient in practice, for it is possible only on very rare occasions to work accurately with the windows open, not at all in summer on account of night noises — the noise of crickets and other insects — and in the winter only when there is but the slightest wind; and further, but few rooms have sufficient window surface to produce the desired absorption. It is necessary, therefore, to work with cushions, but to express the results in open-window units.

Turning now to the unit into which the results are to be transformed, an especially quiet winter night was taken to determine whether the absorbing power of open windows is proportional to the area. A test of the absorbing power of seven windows, each 1.10 meters wide, when opened .20, .40, and .80 meter, gave results that are plotted in the diagram. The points, by falling in a straight line, show that, at least for moderate breadths, the absorbing power of open windows, as of cushions, is accurately proportional to the area. Experiments in several rooms especially convenient for the purpose determined the absorbing power of the cushions to be .80 of that of an equal area of open windows. These cushions were of hair, covered with canvas and light damask. "Elastic Felt" cushions having been used during an investigation in a New York church, it was necessary on returning to Cambridge to determine their absorbing power. This was accomplished through the courtesy of the manufacturers, Messrs. Sperry & Beale, of New York, and the absorbing power was found to be .73 of open-window

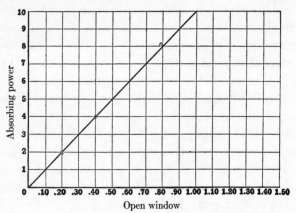

Fig. 7. The absorbing power of open windows plotted against the areas of the openings, showing them to be proportional.

units — an interesting figure, since these cushions are of frequent use and of standard character.

Hereafter all results, though ordinarily obtained by means of cushions, will be expressed in terms of the absorbing power of open windows — a unit as permanent, universally accessible, and as nearly absolute as possible. In these units the total absorbing power of the walls, ceiling, floor, windows and chairs in the lecture-room of the Fogg Museum is 75.5.

Next in order is the apportionment of the total absorbing power among the various components of the structure. Let s_1 be the area of the plaster on tile, and a_1 its absorbing power per square meter; s_2 and a_2 the corresponding values for the plaster on wire lath; s_3 and a_3 for window surface, etc. Then

$$a_1 s_1 + a_2 s_2 + a_3 s_3 + a_4 s_4, \text{ etc. } = 75.5,$$

s_1, s_2, s_3, etc., are known, and a_1, a_2, a_3, etc. — the coefficients of absorption — are unknown, and are being sought. Similar equations may be obtained for other rooms in which the proportions of wall-surface of the various kinds are greatly different, until there are as many equations as there are unknown quantities. It is then possible by elimination to determine the absorbing power of the various materials used in construction.

Through the kindness of Professor Goodale, an excellent opportunity for securing some fundamentally interesting data was afforded by the new Botanical Laboratory and Greenhouse recently given to the University. These rooms — the office, the laboratory and the greenhouse — were exclusively finished in hard-pine sheathing, glass, and cement; the three rooms, fortunately, combined the three materials in very different proportions. They and the constant-temperature room in the Physical Laboratory — the latter being almost wholly of brick and cement — gave the following data:

	Area of Hard Pine Sheathing	Area of Glass	Area of Brick and Cement	Combined Absorbing Power
Office........................	127.0	7	0	8.37
Laboratory...................	84.8	6	30	5.14
Greenhouse..................	12.7	80	85	4.64
Constant-temperature room.....	2.1	0	124	3.08

This table gives for the three components the following coefficients of absorption: hard pine sheathing .058, glass .024, brick set in cement .023.

APPROXIMATE SOLUTION

In the preceding paper it was shown that the duration of the residual sound in a particular room was proportional inversely to the absorbing power of the bounding walls and the contained material, the law being expressed closely by the formula $(a + x)t = k$, the formula of a displaced rectangular hyperbola. In the present paper it is proposed to show that this formula is general, and applicable to any room; that in adapting it to different rooms it is only necessary to change the value of the constant of inverse proportionality k; that k is in turn proportional to the volume of the room, being equal to about .171V in the present experiments, but dependent on the initial intensity of the sound; and finally, that by substituting the value of k thus determined, and also the

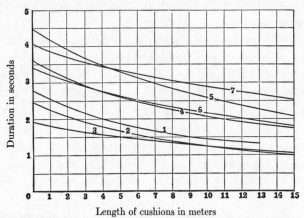

Fig. 8. Curves showing the relation of the duration of the residual sound to the added absorbing material, — rooms 1 to 7.

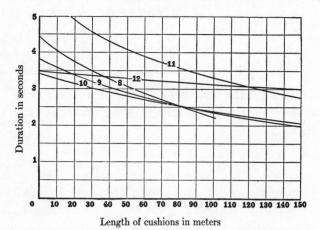

Fig. 9. Curves showing the relation of the duration of the residual sound to the added absorbing material, — rooms 8 to 12.

value of a, the absorbing power of the walls, and of x, the absorbing power of the furniture and audience, it is possible to calculate in advance of construction the duration of audibility of the residual sound.

The truth of the first proposition — the general applicability of the hyperbolic law of inverse proportionality — can be satisfactorily shown by a condensed statement of the results obtained from data collected early in the investigation. These observations were made in rooms varying extremely in size and shape, from a small committee-room to a theatre having a seating capacity for nearly fifteen hundred. Figures 8 and 9 give the curves experimentally determined, the duration of audibility of the residual

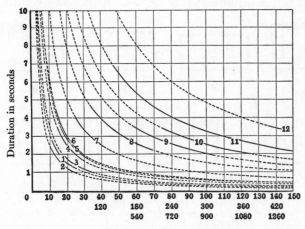

Fig. 10. The curves of Figs. 8 and 9 entered as parts of their corresponding rectangular hyperbolas. Three scales are employed for the volumes, by groups 1–7, 8–11, and 12.

sound being plotted against running meters of cushions. Two diagrams are given in order to employ a smaller scale for the larger rooms, this scale being one-tenth the other; and even in this way there is shown but one-quarter of the curve actually obtained in rooms numbered 11 and 12, the Fogg Art Museum lecture-room and Sanders Theatre. In Fig. 10 each curve is entered as a part of its corresponding hyperbola referred to its asymptotes as axes. In this case three scales are employed in order to show the details more clearly, the results obtained in rooms 1 to 7 on one scale, 8 to 11 on another, and 12 on a third, the three scales being proportional to one, three and nine. The continuous portions of the curves show the parts determined experimentally. Even with the scale thus changed only a very small portion of the experimentally determined parts of curves 11 and 12 are shown. Figures 11 to 16, inclusive, all drawn to the same scale, show the great variation in size and shape of the rooms tested; and the accompanying notes give for each the maximum departure and average departure of the curve, experimentally determined, from the nearest true hyperbola.

1. Committee-room, University Hall; plaster on wood lath, wood dado; volume, 65 cubic meters; original duration of residual sound before the introduction of any cushions, 2.82 seconds; maxi-

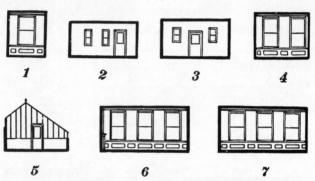

Fig. 11. 1. Committee-room. 2. Laboratory, Botanic Gardens. 3. Office, Botanic Gardens. 4. Recorder's Office. 5. Greenhouse. 6. Dean's Room. 7. Clerk's Room.

mum departure of experimentally determined curve from the nearest hyperbola, .09 second; average departure, .03 second.

2. Laboratory, Botanic Gardens of Harvard University; hard pine walls and ceiling, cement floor; volume, 82 cubic meters; original duration of the residual sound, 2.39 seconds; maximum departure from hyperbola, .09 second; average departure, .02 second.

3. Office, Botanic Gardens; hard pine walls, ceiling and floor; volume, 99 cubic meters; original duration of residual sound, 1.91 seconds; maximum departure from hyperbola, .01 second; average departure, .00 second.

4. Recorder's Office, University Hall; plaster on wood lath, wood dado; volume, 102 cubic meters; original duration of residual sound, 3.68 seconds; maximum departure from hyperbola, .10 second; average departure, .04 second.

5. Greenhouse, Botanic Gardens; glass roof and sides, cement floor; volume, 134 cubic meters; original duration of residual

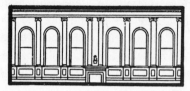

FIG. 12. Faculty-room.

sound, 4.40 seconds; maximum departure from hyperbola, .08 second; average departure, .03 second.

6. Dean's Room, University Hall; plaster on wood lath, wood dado; volume, 166 cubic meters; original duration of residual

FIG. 13. Lecture-room.

sound, 3.38 seconds; maximum departure from hyperbola, .06 second; average departure, .01 second.

7. Clerk's Room, University Hall; plaster on wood lath, wood dado; volume, 221 cubic meters; original duration of residual

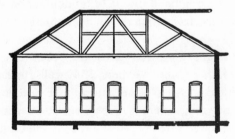

FIG. 14. Laboratory.

sound, 4.10 seconds; maximum departure from hyperbola, .10 second; average departure, .02 second.

8. Faculty-room, University Hall; plaster on wood lath, wood dado; volume, 1,480 cubic meters; original duration of residual sound, 7.04 seconds; maximum departure from hyperbola, .18 second; average departure, .08 second.

9. Lecture-room, Room 1, Jefferson Physical Laboratory; brick walls, plaster on wood lath ceiling; furnished; volume, 1,630 cubic meters; original duration of residual sound, 3.91

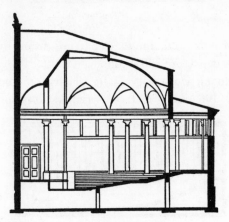

Fig. 15. Lecture-room.

seconds; maximum departure from hyperbola, .10 second; average departure, .04 second.

10. Large Laboratory, Room 41, Jefferson Physical Laboratory; brick walls, plaster on wood lath ceiling; furnished; volume, 1,960 cubic meters; original duration of residual sound, 3.40 seconds; maximum departure from hyperbola, .03 second; average departure, .01 second.

11. Lecture-room, Fogg Art Museum; plaster on tile walls, plaster on wire-lath ceiling; volume, 2,740 cubic meters; original duration of residual sound, 5.61 seconds; maximum departure from hyperbola, .04 second; average departure, .02 second. The experiments in this room were carried so far that the original duration of residual sound of 5.61 seconds was reduced to .75 second.

12. Sanders Theatre; plaster on wood lath, but with a great deal of hard-wood sheathing used in the interior finish; volume, 9,300 cubic meters; original duration of residual sound, 3.42

seconds; maximum departure from hyperbola, .07 second; average departure, .02 second.

It thus appears that the hyperbolic law of inverse proportionality holds under extremely diverse conditions in regard to the size, shape and material of the room. And as the cushions used in the calibration were placed about quite at random, it also appears that in rooms small or large, with high or low ceiling, with flat or curved

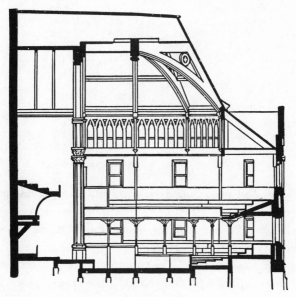

FIG. 16.   Sanders Theatre.

walls or ceiling, even in rooms with galleries, the cushions, wherever placed — out from under the gallery, under, or in the gallery — are nearly equally efficacious as absorbents. This merely means, however, that the efficacy of an absorbent is independent of its position when the problem under consideration is that of reverberation, and that the sound, dispersed by regular and irregular reflection and by diffraction, is of nearly the same intensity at all parts of the room soon after the source has ceased; and it will be the object of a subsequent paper to show that in respect to the initial distribution of the sound, and also in respect to discrete echoes, the position of the absorbent is a matter of prime importance.

Having shown that the hyperbolic law is a general one, interest centers in the parameter, k, the constant for any one room, but varying from room to room, as the following table shows:

Room	Volume	Absorbing Power of Walls, etc., = a	Parameter k
1. Committee-room, University Hall...	65	4.76	13.6
2. Laboratory, Botanic Gardens.......	82	4.65	11.1
3. Office, Botanic Gardens...........	99	8.08	15.4
4. Recorder's Office.................	102	5.91	21.8
5. Greenhouse, Botanic Gardens......	134	5.87	25.8
6. Dean's Room.....................	166	7.50	25.4
7. Clerk's Room....................	221	10.6	43.5
8. Faculty-room....................	1,480	34.5	243.0
9. Lecture-room, Jefferson Physical Laboratory, 1.....................	1,630	69.0	270.0
10. Laboratory, Jefferson Physical Laboratory, 41.....................	1,960	101.0	345.0
11. Fogg Lecture-room..............	2,740	75.0	425.0
12. Sanders Theatre.................	9,300	465.0	1,590.0

The values of the absorbing power, a, and the parameter, k, are here expressed, not in terms of the cushions actually used in the experiments, but in terms of the open-window units, shown to be preferable in the preceding article.

In the diagram, Figure 17, the values of k are plotted against the corresponding volumes of the rooms; here again three different scales are employed in order to magnify the results obtained in the smaller rooms. The resulting straight line shows that the value of k is proportional to the volume of the room, and it is to be observed that the largest room was nearly one hundred and fifty times larger than the smallest. By measurements of the coördinates of the line, or by averaging the results found in calculating $\dfrac{k}{V}$ for all the rooms it appears that $k = .171V$. The physical significance of this numerical magnitude .171 will be explained later.

This simple relationship between the value of k and the volume of the room — the rooms tested varying so greatly in size and shape — affords additional proof, by a rather delicate test, of the accuracy of the method of experimenting, for it shows that the experimentally determined curves approximate not merely to hyperbolas but to a systematic family of hyperbolas. It also furnishes a more pleasing prospect, for the laborious handling of cushions will be unnecessary. A single experiment in a room and a knowledge of the volume of the room will furnish sufficient data for the calculation of the absorbing power of its components. Conversely, a knowledge of the volume of a room and of the coefficients of absorption of its various components, including the audience for which it is designed, will enable one to calculate in advance of construction the duration of audibility of the residual sound, which measures

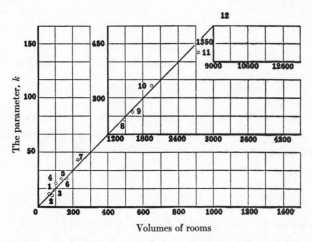

Fig. 17. The parameter, k, plotted against the volumes of the rooms, showing the two proportional.

that acoustical property of a room commonly called reverberation. Therefore, this phase of the problem is solved to a first approximation.

The explanation of the fact that k is proportional to V is found in the following reasoning. Consider two rooms, constructed of exactly the same materials, similar in relative proportions, but one larger than the other. The rooms being empty, x, the absorbing power of the contained material, is zero, and we have $a' t' = k'$ and $a'' t'' = k''$. Since the rooms are constructed of the same materials the coefficients of absorption are the same, so that a' and a'' are proportional to the surfaces of the rooms, that is, to the squares of the linear dimensions. Also, the residual sound is diminished a certain percentage at each reflection, and the more frequent these reflections are the shorter is the duration of its audibility; whence t' and t'' are inversely proportional to the frequency of the reflections, and hence directly proportional to the linear dimensions. Therefore, k' and k'', which are equal to $a' t'$ and $a'' t''$, are proportional to the cubes of the linear dimensions, and hence to the volumes of the rooms.

Further, when the shape of the room varies, the volume remaining the same, the number of reflections per second will vary. Therefore, k is a function not merely of the volume, but also of the shape of the room. But that it is only a slightly varying function, comparatively, of the shape of the room for practical cases, is shown by the fact that the points fall so near the straight line that averages the values of the ratio $\dfrac{k}{V}$.

The value of k is also a function of the initial intensity of the sound; but the consideration of this element will be taken up in a following paper.

RATE OF DECAY OF RESIDUAL SOUND

In a subsequent discussion of the interference of sound it will be shown by photographs that the residual sound at any one point in the room as it dies away passes through maxima and minima, in many cases beginning to rise in intensity immediately after the source has ceased; and that these maxima and minima succeed each other in a far from simple manner as the interference system shifts. On this account it is quite impossible to use any of the numerous direct methods of measuring sound in experiments on reverberation. Or, rather, if such methods were used the results would be a mass of data extremely difficult to interpret. It was for this reason that attempts in this direction were abandoned early in the investigation, and the method already described adopted. In addition to the fact that this method only is feasible, it has the advantage of making the measurements directly in terms of those units with which one is here concerned — the minimum audible intensity. It is now proposed to extend this method to the determination of the rate of decay of the average intensity of sound in the room, and to the determination of the intensity of the initial sound, and thence to the determination of the mean free path between reflections, — all in preparation for the more exact solution of the problem.

The first careful experiment on the absolute rate of decay was in the lecture-room of the Boston Public Library, a large room,

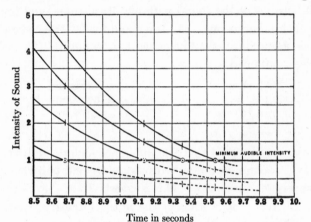

FIG. 18.   Decay of sound in the lecture-room of the Boston Public Library from the initial sound of one, two, three, and four organ pipes, showing only the last second.

finished, with the exception of the platform, in material of very slight absorbing power — tile ceiling, plaster on tile walls, and polished cement floor.[1]   The reverberation was very great, 8.69 seconds.   On the platform were placed four organ pipes, all of the same pitch, each on its own tank or wind supply, and each having its own electro-pneumatic valve.   All these valves, however, were connected to one chronograph, key, and battery, so that one, two, three, or all the pipes, might be started and stopped at once, and when less than four were in use any desired combination could be made.   One pipe was sounded and the duration of audibility of the residual sound determined, of course, as always in these experiments, by repeated observations.   The experiment was then made

[1] Terrazzo cement floor.

with two organ pipes instead of one; then with three pipes; and, finally, with four. The whole series was then repeated, but beginning with a different pipe and combining different pipes for the two and three pipe sets. In this way the series was repeated four times, the combinations being so made that each pipe was given an equal weight in the determination of the duration of audibility of the residual sound under the four different conditions. It is safe to assume that with experiments conducted in this manner the average initial intensities of the sound with one, two, three, and four pipes were to each other as one, two, three and four. The corresponding durations of audibility shall be called t_1, t_2, t_3 and t_4. The following results were obtained:

$$t_1 = 8.69 \text{ seconds} \qquad t_2 - t_1 = .45 \text{ second}$$
$$t_2 = 9.14 \text{ " } \qquad t_3 - t_1 = .67 \text{ "}$$
$$t_3 = 9.36 \text{ " } \qquad t_4 - t_1 = .86 \text{ "}$$
$$t_4 = 9.55 \text{ "}$$

It is first to be observed that the difference for one and two organ pipes, .45, is, within two-hundredths of a second, half that for one and four organ pipes, .86. This suggests that the difference is proportional to the logarithm of the initial intensity; and further inspection shows that the intermediate result with three organ pipes, .67, is even more nearly, in fact well within a hundredth of a second, proportional to the logarithm of three. This reënforces the very natural conception that however much the residual sound at any one point in the room may fluctuate, passing through maxima and minima, the average intensity of sound in the room dies away logarithmically. Thus, if one plots the last part of the residual sound — that which remains after eight seconds have elapsed — on the assumption that the intensity of the sound at any instant is proportional to the initial intensity, the result will be as shown in the diagram, Fig. 18. The point at which the diminishing sound crosses the line of minimum audibility in each of the four cases is known, the corresponding ordinates of the other curves being multiples or submultiples in proportion to the initial intensity. The results are obviously logarithmic.

Let I_1 be the average intensity of the steady sound in the room when the single organ pipe is sounding, i the intensity at any instant during the decay, say t seconds after the pipe has ceased, then $-\dfrac{di}{dt}$ will be the rate of decay of the sound, and since the absorption of sound is proportional to the intensity

$$-\frac{di}{dt} = Ai, \text{ where } A \text{ is the constant of proportionality,}$$

the ratio of the rate of decay of the residual sound to the intensity at the instant.

$$- log_e i + C = At,$$

a result that is in accord with the above experiments. The constant of integration C may be determined by the fact that when t is zero i is equal to I_1; whence

$$C = log_e I_1, \text{ and the above equation becomes}$$
$$log_e \frac{I_1}{i} = At.$$

At the instant of minimum audibility t is equal to t_1, the whole duration of the residual sound, and i is equal to i', — as the intensity of the least audible sound will hereafter be denoted. Therefore

$$log_e \frac{I_1}{i'} = At_1.$$

This applied to the experiment with two, three and four pipes gives similar equations of the form

$$log_e \frac{nI_1}{i'} = At_n,$$

where n is the number of organ pipes in use. By the elimination of $\dfrac{I_1}{i'}$ from these equations by pairing the first with each of the others,

$$A = \frac{log_e 2}{t_2 - t_1} = 1.54,$$

$$A = \frac{log_e 3}{t_3 - t_1} = 1.62,$$

$$A = \frac{log_e 4}{t_4 - t_1} = 1.61,$$

$$A(\text{average}) = \overline{1.59},$$

where A is the ratio between the rate of decay and the average intensity at any instant.

It is possible also to determine the initial intensity, I_1, in terms of the minimum audible intensity, i'.

$$log_e \frac{I_1}{i'} = At_1,$$
$$I_1 = i' \, log_e^{-1} \, At_1 = i' \, log_e^{-1} \, (1.59 \times 8.69) = 1,000,000 \, i'.$$

With this value of the initial intensity it is possible to calculate the intensity i of the residual sound at any instant during the decay, by the formula $\qquad log_e I_1 - log_e i = At,$

and the result when plotted is shown in Figure 19, the unit of intensity being minimum audibility.

A practical trial early in the year had shown that it would be impossible to use this lecture-room as an auditorium, and the ex-

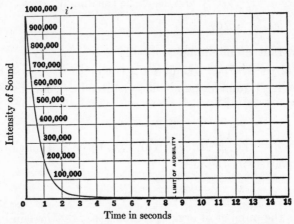

FIG. 19. Decay of sound in the lecture-room of the Boston Public Library beginning immediately after the cessation of one organ pipe.

periments described above, with others, were in anticipation of changes designed to remedy the difficulty. Hair felt, in considerable quantities, was placed on the rear wall. The experiments with the four organ pipes were then repeated and the following results were obtained:

$t_1 = 3.65$	$t_2 - t_1 = .20$	$\therefore A = 3.41$
$t_2 = 3.85$	$t_3 - t_1 = .31$	$\therefore A = 3.54$
$t_3 = 3.96$	$t_4 - t_1 = .42$	$\therefore A = \underline{3.29}$
$t_4 = 4.07$		$A = \overline{3.41}$ (average)
$I_1 = 250,000 \, i'$		

A few nights later the apparatus was moved down to the attendant's reception-room near the main entrance — a small room but similar in proportions to the lecture-room. Here a careful experiment extending over several nights was carried on, and it gave the following results:

$$t_1 = 4.01 \qquad t_2 - t_1 = .19 \qquad \therefore A = 3.65$$
$$t_2 = 4.20 \qquad t_3 - t_1 = .28 \qquad \therefore A = 3.90$$
$$t_3 = 4.29 \qquad t_4 - t_1 = .37 \qquad \therefore A = \underline{3.75}$$
$$t_4 = 4.38 \qquad\qquad\qquad\qquad A = 3.76 \text{ (average)}$$
$$I_1 = 3,800,000 \; i'$$

The first interest lies in an attempt to connect the rate of decay, obtained by means of the four organ pipe experiments, with the absolute coefficient of absorption of the walls, obtained by the experiments with the open and closed windows; and to this end recourse will be had to what shall here be called "the mean free path between reflections." The residual sound in its process of decay travels across the room from wall to wall, or ceiling, or floor, in all conceivable directions; some paths are the whole length of the room, some even longer, from one corner to the opposite, but in the main the free path between reflections is less, becoming even infinitesimally small at an angle or a corner. Between the two or three hundred reflections that occur during its audibility the residual sound establishes an average distance between reflections that depends merely on the dimensions of the room, and may be called "its mean free path."

$$a = \frac{.171 \, V}{t_1}$$

is the absorbing power of the room, measured in open-window units.
Let

$$s \; = \text{surface.}$$
$$V \; = \text{volume.}$$
$$A \; = \text{rate of decay of the sound.}$$
$$v \; = \text{velocity of sound, 342 m. per second at 20 degrees C.}$$
$$p \; = \text{length of the mean free path between reflections.}$$

Whence $\dfrac{v}{p}$ = the average number of reflections per second, and

$\dfrac{a}{s}$ is the fraction absorbed at each reflection, $\dfrac{av}{sp} = A,$

and $p = \dfrac{av}{As} = \dfrac{v.171\ V}{As\ t_1}$, whence may be calculated the mean free path, p.

	V	A	s	t	p
Boston Public Library Lecture-room, bare	2,140.0	1.59	1,160	8.69	7.8
" " " " with felt . .	2,140.0	3.41	1,160	3.65	8.8
" " " Attendant's Room	63.8	3.76	108	4.01	2.27

The length of the mean free path in the lecture-room, bare or draped, ought to be the same, for the felt was placed out from the wall at a distance imperceptibly small in comparison with the dimensions of the room; but 7.8 and 8.8 differ more than the experimental errors justify. Again, the attendant's room had very nearly the same relative proportions as the lecture-room (about 2 : 3 : 6), but each linear dimension reduced in the ratio 3.22 : 1. The mean free path, obviously, should be in the same ratio; but when the mean free path in the attendant's room, 2.27, is multiplied by 3.22 it gives 7.35, departing again from the other values, 7.8 and 8.8, more than experimental errors justify. The explanation of this is to be found in the fact that the initial intensity of the sound in the rooms for the determination of t_1 was not the same but had the values respectively, 1,000,000 i', 250,000 i' and 3,800,000 i'. Since t_1 has been shown proportional to the logarithms of the initial intensities, these three numbers, 7.8, 8.8 and 7.35, may be corrected in an obvious manner, and reduced to the comparable values they would have had if the initial intensity had been the same in all three cases. The results of this reduction are 7.8, 8.0 and 8.0, a satisfactory agreement.

The length of the mean free path is, therefore, as was to be expected, proportional to the linear dimensions of the room, and such a comparison is interesting. There is no more reason, however, for comparing it with one dimension than another. Moreover, most rooms in regard to which the inquiry might be made are too irregular in shape to admit of any one actual distance being taken as standard. Thus, in a semicircular room, still more in a horseshoe-shaped room such as the common theatre, it is indeterminable what should be called the breadth or what the length. On account, therefore, of the complicated nature of practical conditions one is forced to the adoption of an ideal dimension, the cube root of the volume, $V^{1/3}$, the length of one side of a cubical room of the same capacity. The above data give as the ratio of $\dfrac{p}{V^{1/3}}$ the value, .62.

It now becomes possible to present the subject by exact analysis, and free from approximations; but before doing so it will be well to review from this new standpoint that which has already been done.

It was obvious from the beginning, even in deducing the hyperbolic law, that some account should be taken of the reduction in the initial intensity of the sound as more and more absorbing material was brought into the room, even when the source of sound remained unchanged. Thus each succeeding value of the duration of the residual sound was less as more and more absorbing material was brought into the room, not merely because the rate of decay was greater, but also because the initial intensity was less. Had the initial intensity in some way been kept up to the same value throughout the series, the resulting curve would have been an exact hyperbola. As it was, however, the curve sloped a little more rapidly on account of the additional reduction in the duration arising from the reduction in initial intensity of the sound. At the time, there was no way to make allowance for this. That it was a very small error, however, is shown by the fact that the departures from the true hyperbola that were tabulated are so small.

Turning now to the parameter, k, it is evident that this also was an approximation, though a close one. In the first place, as just explained, the experimental curve of calibration sloped a little more rapidly than the true hyperbola. It follows that the nearest hyperbola fitting the actual experimental results was always of a little too small parameter. Further, k depended not merely on the uniformity of the initial intensity during the calibration of the room, but also on the absolute value of this intensity. Thus, $k = at_1$, and t_1 is in turn proportional to the logarithm of the initial intensity. Therefore in order to fully define k we must adopt some standard of initial intensity. For this purpose we shall hereafter take as the standard condition in initial intensity, $I = 1,000,000 \, i'$, $(I = 10^6 \, i')$, where i' is the minimum audible intensity, as this is the nearest round number to the average intensity prevailing during these experiments. If, therefore, during the preceding experiments the initial intensity was above the standard, the value deduced for k would be a little high, if below standard, a little low. This variation of the parameter, k, would be slight ordinarily, for k is proportional to the logarithm, not directly to the value of the initial intensity. Slight ordinarily, but not always. Attention was first directed to its practical importance early in the whole investigation by an experiment in the dining-room of Memorial Hall — a very large room of 17,000 cubic meters capacity. During some experiments in Sanders Theatre the organ pipe was moved across to this dining-room, and an experiment begun. The reverberation was of very short duration, although it would have been long had the initial intensity been standard, for in rooms constructed of similar materials the reverberation is approximately proportional to the cube roots of the volumes. There was no opportunity to carry the experiment farther than to observe the fact that the duration was surprisingly short, for the frightened appearance of the women from the sleeping-rooms at the top of the hall put an end to the experiment. Finally, k is a function not merely of the volume but also of the shape of the room; that is to say, of the mean free path, as has already been explained.

It was early recognized that with a constant source the average intensity of the sound in different rooms varies with variations in size and construction, and that proper allowance should be made therefor. The above results call renewed attention to this, and point the way. In the following paper the more exact analysis will be given and applied.

EXACT SOLUTION

THE present paper will carry forward the more exact analysis proposed in the last paper.

For the sake of reference the nomenclature so far introduced is here tabulated.

t	= time after the source has ceased up to any instant whatever during the decay of the sound.
t', t'', t'''	= duration of the residual sound, the accents indicating a changed condition in the room such as the introduction or removal of some absorbent, the presence of an audience, or the opening of a window.
$t_1, t_2, \ldots t_n$	= whole duration of the residual sound, the subscripts indicating the number of organ pipes used.
T	= duration of the residual sound in a room when the initial intensity has been standard.
i	= intensity of the residual sound at any instant.
i'	= intensity of minimum audibility.
$I_1, I_2, \ldots I_n$	= intensity of sound in the room just as the organ pipe or pipes stop, the subscripts indicating number of pipes.
I	= standard initial intensity arbitrarily adopted, $I = 1,000,000\ i'$.
w	= absorbing power of the open windows, minus their absorbing power when closed = area $(1 - .024)$.
a	= absorbing power of the room.
$a_1, a_2, \ldots a_n$	= coefficients of absorption of the various components of the wall-surface.
s	= area of wall (and floor) surface in square meters.
$s_1, s_2, \ldots s_n$	= area of the various components of the wall-surface.
V	= volume of the room in cubic meters.
k	= hyperbolic parameter of any room.
K	= ratio of the parameter to the volume, $aT = k = KV$.
A	= rate of decay of the sound.
p	= length of mean free path between reflections.
v	= velocity of sound, 342 m. per second at 20° C.

Let E denote the rate of emission of energy from the single organ pipe.

$$\frac{p}{v} = \text{the average interval of time between reflections.}$$

$$\frac{p}{v}E = \text{amount of energy emitted during this interval.}$$

$$\frac{p}{v}E\left(1 - \frac{a}{s}\right) = \text{amount of energy left after the first reflection.}$$

$$\frac{p}{v}E\left(1 - \frac{a}{s}\right)^2 = \text{amount of energy left after the second reflection, etc.}$$

If the organ pipe continues to sound, the energy in the room continues to accumulate, at first rapidly, afterwards more and more slowly, and finally reaches a practically steady condition. Two points are here interesting, — the time required for the sound to reach a practically steady condition (for in the experiments the organ pipes ought to be sounded at least this long), and second, the intensity of the sound in the steady and final condition. At any instant, the total energy in the room is that of the sound just issuing from the pipe, not having suffered any reflection, plus the energy of that which has suffered one reflection, that which has suffered two, that which has suffered three, and so on back to that which first issued from the pipe, as:

$$\frac{p}{v} E \left[1 + \left(1 - \frac{a}{s} \right) + \left(1 - \frac{a}{s} \right)^2 + \left(1 - \frac{a}{s} \right)^3 + \ldots \left(1 - \frac{a}{s} \right)^n \right],$$

where n is the number of reflections suffered by the sound that first issued from the pipe, and is equal to the length of time the pipe was blown divided by the average interval of time between reflections. The above series, which is an ordinary geometric progression, may be written

$$\frac{p}{v} E \frac{1 - \left(1 - \frac{a}{s} \right)^n}{1 - \left(1 - \frac{a}{s} \right)} ; \tag{1}$$

$\frac{a}{s}$ is by nature positive and less than unity. If n is very large or if $\left(1 - \frac{a}{s} \right)$ is small this may be written

$$\frac{pEs}{va} = \text{the total energy in the room in the steady condition.} \tag{2}$$

$$I_1 = \frac{pEs}{avV} \tag{3}$$

is the average intensity of sound in the room as the organ pipe stops. Substituting in this equation the values of a and p already found,

$$a = \frac{KV}{T}, \tag{4}$$

and

$$p = \frac{va}{sA} = \frac{vKV}{sAT}, \tag{5}$$

we have $\qquad I_1 = \dfrac{vKV}{sAT} \cdot \dfrac{T}{KV} \cdot \dfrac{Es}{vV} = \dfrac{E}{AV}.$ $\qquad$ (6)

Also
$$I_1 = log_e^{-1} At_1,$$ (7)

whence
$$E = VA \, log_e^{-1} At_1,$$ (8)

where the unit of energy is the energy of minimum audibility in a cubic meter of air.

It remains to determine K and a. To this end the four organ pipe experiments must be made in a room with the windows closed and with them open, and the values of A' and A'' determined. The following analysis then becomes available:

$$a = \frac{KV}{T'}, \text{ and } a + w = \frac{KV}{T''},$$

whence
$$\frac{a}{a+w} = \frac{T''}{T'}.$$

For standard conditions in regard to initial intensity
$$A' \, T' = A'' \, T'' = log_e I = log_e (10^6) = 13.8,$$
$$\frac{T''}{T'} = \frac{A'}{A''}, \text{ and } T' = \frac{13.8}{A'}.$$

Substituting these values,
$$\frac{a}{a+w} = \frac{A'}{A''}, K = \frac{aT'}{V} = \frac{a \, 13.8}{A' \, V};$$

whence
$$a = \frac{A' \, w}{A'' - A'},$$ (9)

and
$$K = \frac{13.8w}{V \, (A'' - A')}.$$ (10)

Or if K has been determined (9) may be written
$$a = \frac{A' \, KV}{13.8},$$ (11)

a useful form of the equation.

From equation (1) and (2) we may calculate the rate of growth of sound in the room as it approaches the final steady condition.

Thus, dividing (1) by (2), the result, $1 - \left(1 - \dfrac{a}{s}\right)^n$, gives the intensity at any instant $n\dfrac{p}{v}$ seconds after the sound has started, in terms of the final steady intensity. Of all the rooms so far experimented on, the growth of the sound was slowest in the lecture-room of the Boston Public Library in its unfurnished condition. For this room $\dfrac{a}{s} = .037$, and $p = 8.0$ meters. The following table shows the growth of the sound in this room, and the corresponding number of reflections which the sound that first issued from the pipe had undergone.

LECTURE-ROOM, BOSTON PUBLIC LIBRARY

n	Time	Average Intensity	n	Time	Average Intensity
1	.02	.04	30	.69	.68
5	.11	.17	40	.92	.78
10	.23	.31	50	1.15	.85
15	.34	.43	100	2.30	.98
20	.46	.53	150	3.45	.997
			∞	∞	1.00

It thus appears that in this particular room the organ pipe must sound for about three seconds in order that the average intensity of the sound may get within ninety-nine per cent of its final steady value. As throughout this work we are concerned only with the logarithm of the initial intensity, ninety-nine per cent of the steady condition is abundantly near. This consideration — the necessary length of time the organ pipe should sound — is carefully regarded throughout these experiments. It varies from room to room, being greater in large rooms, and less in rooms of great absorbing power.

To determine the value of E, the rate of emission of sound by the pipe, formula (8), $E = VA \, log_e^{-1} \, At_1$, is available. It is here to be observed that as this involves the antilogarithm of At_1 these quantities must be determined with the greatest possible accuracy. The first essential to this end is the choice of an appropriate room. Without giving the argument in detail here, it leads to this, that the best rooms in which to experiment are those that are large in volume and have little absorbing power. In fact, for this purpose, small rooms are almost useless, but the accuracy of the result increases rapidly with an increase in size or a decrease in absorbing power. On this account the lecture-room of the Boston Public Library in its unfurnished condition was by far the best for this determination of all the available rooms. Inserting the numerical magnitudes obtained in this room in the equation,

$$E = VA \, log_e^{-1} \, At_1 = 2,140 \times 1.59 \, log_e^{-1} \, (1.59 \times 8.69) = 3,400,000,000.$$

If the observations in the same room after the introduction of the felt, already referred to, are used in the equation the resulting value of E is 3,200,000,000. The agreement between the two is merely fortunate, for the second conditions were very inferior to the first, and but little reliance should be placed on it. In fact, in both results the second figures, 4 and 2, are doubtful, and the round number, 3,000,000,000, will be used. It is sufficiently accurate.

The next equation of interest is that giving the value of K, number (10). It contains the expression, $A'' - A'$, the difference between the rates of decay with the windows open and with them closed; A'' and A' depend linearly on the difference in duration of the residual sound with four organ pipes and with one, and as both sets of differences are at best small, it is evident that these experiments also must be conducted with the utmost care and under the best conditions. The best conditions would be in rooms that are large, that have small absorbing power, and that afford window area sufficient to about double the absorbing power of the room. Practically this would be in large rooms that are of tile, brick, or cement walls, ceiling and floor, and have an available window area equal to about one-thirtieth of the total area.

The lobby of the Fogg Art Museum, although rather small, best satisfied the desired conditions. Sixteen organ pipes were used, arranged four on each air tank and, therefore, near together. Thus arranged, the sixteen pipes had 7.6 times the intensity of one, as determined by a subsequent experiment in the Physical Laboratory. The following results were obtained:

$$A' = \frac{log_e 7.6}{t'_{16} - t'_1} = \frac{log_e 7.6}{5.26 - 4.59} = 3.0,$$

and

$$A'' = \frac{log_e 7.6}{3.43 - 3.00} = 4.7.$$

$$K = \frac{13.8w}{V(A'' - A')} = \frac{13.8 \times 1.85}{96 \times 1.7} = .156.$$

Here, however, it is easy to show by trial that errors of only one-hundredth of a second in the four determinations of the duration of the residual sound would, if additive, give a total error of twenty per cent in the result.

It is impossible, especially with open windows, to time with an accuracy of more than one-hundredth of a second, and, therefore, this formula,

$$K = \frac{13.8w}{V(A'' - A')},$$

while analytically exact and attractive in its simplicity, is practically unserviceable on account of the sensitive manner in which the observations enter into the calculations.

The following analysis, however, results in an equation much more forbidding in appearance, it is true, but vastly better practically, for it involves the data of difficult determination only logarithmically, and then only as part of a comparatively small correcting term. For the room with the windows closed:

$$A' \, t'_1 = log_e \, I'_1;$$

and for standard conditions in regard to initial intensity

$$A' \, T' = log_e \, I,$$

whence

$$T' \;\; = t'_1 - \frac{1}{A'} \, log_e \, \frac{I'_1}{I};$$

$$T' \, a = KV,$$

hence

$$KV \;\; = t'_1 \, a - \frac{a}{A'} \, log_e \, \frac{I'_1}{I};$$

and similar steps for the same room with the windows open give

$$KV = t''_1 \, (a + w) - \frac{(a + w)}{A''} \, log_e \, \frac{I''_1}{I}.$$

Multiplying the first of the last two equations by t''_1, and the second by t'_1,

$$K = \frac{1}{(t'_1 - t''_1) \, V} \left[wt'_1 t''_1 + \left(\frac{at''_1}{A'} \, log_e \, \frac{I'_1}{I} - \frac{(a + w)t'_1}{A''} \, log_e \, \frac{I''_1}{I} \right) \right].$$

By equation (5)

$$\frac{a}{A'} = \frac{sp}{v};$$

and similarly

$$\frac{a + w}{A''} = \frac{sp}{v}.$$

Substituting these values in the above equation,

$$K = \frac{1}{(t'_1 - t''_1)V}\left[wt'_1t''_1 + \frac{sp}{v}\left(t''_1 \, log_e \, \frac{I'_1}{I} - t'_1 \, log_e \, \frac{I''_1}{I}\right)\right]. \quad (12)$$

As an illustration of the application of the last equation, the case of the lobby of the Fogg Art Museum is here worked out at length.

$$t'_1 = 4.59$$
$$t''_1 = 3.00$$
$$V = 96 \text{ cu. m.}$$
$$s = 125 \text{ sq. m.}$$
$$w = 1.86$$
$$a = \frac{.171\,V}{t'_1} = 3.58 \text{ as a first approximation}$$
$$p = 2.8$$
$$I'_1 = \frac{pEs}{vaV} = 8.8 \times 10^{\,6}\,i'$$
$$I''_1 = \frac{pEs}{v(a + w)\,V} = 5.8 \times 10^{6}\,i'$$

Substituting these values in the above equation,

$$K = \frac{1}{152}\left[25.7 + 1.02\,(6.53 - 8.1)\right] = .169 - .010 = .159,$$

where the term .169 is the value of K that would be deduced disregarding the initial intensity of the sound, $-$ .010 is the correction for this, and .159 is the corrected value of K. The magnitude as well as the sign of this correction depends on the intensity of the source of sound, the size of the room and the material of which it is constructed, and the area of the windows opened. This is illustrated in the following table, which is derived from a recalculation of all the rooms in which the open-window experiment has been tried, and which exhibits a fairly large range in these respects:

Room	V	I'_1	w	Uncorrected	Correction	K
Lobby Fogg Museum............	96	8,800,000	1.86	.169	−.010	.159
Lobby Fogg Museum, 16 pipes...	96	67,000,000	1.86	.191	−.027	.164
Jefferson Physical Laboratory 15 .	202	1,700,000	5.10	.164	+.005	.169
Jefferson Physical Laboratory 1 ..	1,630	390,000	12.0	.150	+.017	.167
Jefferson Physical Laboratory 41 .	1,960	300,000	14.6	.137	+.024	.161

Average value of K = .164

The value, K = .164, having been adopted, interest next turns to the determination of the absorbing power, a, of a room. For this purpose we have choice of three equations, two of which have already been deduced, (9) and (11),

$$a = \frac{A' w}{A'' - A'},$$

and

$$a = \frac{A' KV}{13.8},$$

and a third equation may be obtained as follows:
It has been shown that

$$T' = t'_1 - \frac{sp}{va} \log_e \frac{I'_1}{I},$$

and

$$T'a = KV.$$

Therefore

$$at'_1 - \frac{sp}{v} \log_e \frac{I'_1}{I} = KV,$$

and

$$a = \frac{1}{t'_1} \left(KV + \frac{sp}{v} \log_e \frac{I'_1}{I} \right). \tag{13}$$

Of these three equations the first, (9), for reasons already pointed out in regard to a similar equation for K, while rigorously correct, yields a result of great uncertainty on account of its sensitiveness to slight errors in the several determinations of the duration of the residual sound. The second, (11), is very much better than the first, but still not satisfactory in this respect. The third, (13), is wholly satisfactory. It has the same percentage accuracy as t'_1, and the only elements of difficult determination enter logarithmically in a small correcting term.

As an illustration of the application of these equations we may again cite the case of the lobby of the Fogg Art Museum:

by equation (9), $a = \dfrac{3.0 \times 1.86}{4.7 - 3.0} = 3.3;$

by equation (11), $a = \dfrac{3.0 \times .164 \times 96}{13.8} = 3.4;$

by equation (13), $a = \dfrac{1}{4.59} (.164 \times 96 + 1.02 \times log_e 8.8) = 3.8.$

The first two are approximate only, the last, 3.8, is correct, with certainty in regard to the last figure.

There is but one other subject demanding consideration in this way, — the calculation of the absorbing power of objects brought into the room, as cushions, drapery, chairs, and other furniture. This may be approached in two ways, either by means of the rate of decay of the sound and the four organ pipe experiment, or by open-window calibration and a single organ pipe.

Let A''' be the rate of decay when the object is in the room, A' being the rate when the room is empty. Then if a' is the absorbing power of the object:

$$a = \frac{A' \, KV}{13.8},$$

and

$$a + a' = \frac{A''' \, KV}{13.8}.$$

Whence

$$a' = (A''' - A') \frac{KV}{13.8}. \tag{14}$$

Or from the other point of view, equation (13),

$$a = \frac{1}{t'_1} \left(KV + \frac{sp}{v} log_e \frac{I'_1}{I} \right),$$

$$a + a' = \frac{1}{t'''_1} \left(KV + \frac{sp}{v} log_e \frac{I'''_1}{I} \right),$$

whence

$$a' = \frac{KV \, (t'_1 - t'''_1)}{t'_1 t'''_1} - \frac{sp}{v} \left(\frac{1}{t'_1} log_e \frac{I'_1}{I} - \frac{1}{t'''_1} log_e \frac{I'''_1}{I} \right); \tag{15}$$

where I'_1 and I'''_1 are to be calculated as heretofore by a preliminary and approximate estimate of a and a'.

Here also it is easy to show *a priori* that the first equation, (14), while perfectly correct and analytically rigorous, is excessively sensitive to very slight errors of observation, and that on this account equation (15) is decidedly preferable. For example, felt being brought into the lobby of the Fogg Lecture-room and placed on the floor, the values of A''' and t'''_1 were determined to be, respectively, 4.9 and 2.79. Borrowing from the preceding experiment, and substituting in equations (14) and (15) we have

$$a' = (4.9 - 3.0)\,\frac{.164 \times 96}{13.8} = 2.2,$$

$$a' = \frac{.164 \times 96\,(4.59 - 2.79)}{4.59 \times 2.79} - 1.02\left(\frac{1}{4.59}\,log_e\,8.8 - \frac{1}{2.79}\,log_e\,6.1\right) = 2.4,$$

a very satisfactory agreement in view of the extreme sensitiveness of equation (14).

Thus three equations have been deduced, number (12) for the calculation of the parameter, k, (13) for the absorbing power, a, of the wall-surface, and (15) for the absorbing power, a', of introduced material. Each has been verified by other equations analytically rigorous, and developed along very different lines of attack. In each case the agreement was satisfactory, especially in view of the extreme sensitiveness of the equations used as checks.

In the succeeding paper will be deduced, by the method thus established, the coefficients of absorption of the materials that are used ordinarily in the construction and furnishing of an auditorium.

THE ABSORBING POWER OF AN AUDIENCE, AND OTHER DATA

In this paper will be given all the data ordinarily necessary in calculating the reverberation in any auditorium from its plans and specifications. In order to show the degree of confidence to which these data are entitled a very brief account will be given of the experiments by means of which they were obtained. Such an account is especially necessary in the case of the determination of the absorbing power of an audience. This coefficient is, in the nature of things, a factor of every problem, and in a majority of cases it is one of the most important factors; yet it can be determined only through the courtesy of a large number of persons, and even then is attended with difficulty.

The formulas that will be used for the calculation of absorbing power are numbers (13) and (15) in the preceding paper, the correcting terms being at times of considerable importance. The application of these formulas having been illustrated, the whole discussion here will be devoted to the conditions of the experiments and the results obtained.

In every experiment the unavoidable presence of the observer increases the absorbing power. In small rooms, and in large rooms if bare of furniture, the relative increase is considerable, and should always be subtracted from the immediate results of the experiment in order to determine the absorbing power of the room alone. The quantity to be subtracted is constant, provided the same clothes are always worn, and may be determined once for all. For this determination another observer made a set of experiments in a small and otherwise empty room before and after the writer had entered with a duplicate set of apparatus, — air tank, chronograph, and battery. In fact, two persons made independent observations, giving consistently the result that the writer, in the clothes and with the apparatus constantly employed, had an absorbing power of .48 of a unit. For the sake of brevity no further mention will be made of this, but throughout the work this correction is applied wherever necessary.

In the second paper of this series a preliminary calculation was made of the absorbing power of certain wall-surfaces, and the object in so doing was to get an approximate value for the absorbing power of glass. It had been decided that the most convenient unit of absorbing power was a square meter of open window. It is evident, however, that the process of opening a window during the progress of an experiment is merely substituting the absorbing power of the open window for that of the same window closed, — a consideration of possible moment in the nicer development of the subject. This preliminary calculation was in anticipation of and preparation for the more close analysis in the fifth paper. If these coefficients are now calculated, using the corrected formulas of the fifth paper, we arrive at the following results: Cement, and brick set in cement, .025, glass, .027 and wood sheathing, .061.

The experiments in the Boston Public Library gave results that are interesting from several points of view. The total absorbing power of the large lecture-room was found to be 38.9 units distributed as follows: A platform of pine sheathing, exposing a total area of 70 square meters, had an absorbing power of $70 \times .061 = 4.3$; 72 square meters of glass windows had an absorbing power of $72 \times .027 = 1.9$; three large oil paintings, with a total area of 17.4 square meters, had an absorbing power of $17.4 \times .28 = 4.9$; the remainder, 27.8 units, was that of the cement floor, tile ceiling, and plaster on tile walls, in total area 1,095 square meters. This gives as the coefficient of absorption for such construction .0254. A similar calculation of results obtained in the attendant's room in the same building — a room in which the construction of the floor, walls, and ceiling is similar to that in the lecture-room — gives for the value of the coefficient, .0255. The very close agreement of these results, and their agreement with the coefficient, .0251, for cement floor and solid walls of brick set in cement in the constant-temperature room, is satisfactory. However, a far more interesting consideration is the following:

Heretofore in the argument it has been assumed, tacitly, that the total absorption of sound in a room is due to the walls, furniture and audience. There is one other possible absorbent, and only one — the viscosity of the vibrating air. It is now possible to present the argument that led to the conclusion that this, the viscosity of the air throughout the body of the room, is entirely negligible in comparison with the other sources of absorption. These two rooms in the Boston Public Library — the lecture-room and the attendant's room — had, in their bare and unfurnished condition, less absorbing power in the walls than any other rooms of their size yet found. Therefore, if the viscosity of the air is a practical factor it ought to have shown in these two rooms if ever. Fortunately, also, the two rooms differed greatly in size, the volume of one being about thirty-five times that of the other, while the ratio of the areas of the wall-surfaces was about twelve. That part of the absorption due to the walls was proportional to the areas of the walls, and the part due to the viscosity of the air was proportional to the volumes of the rooms. As a matter of fact the experiments in these two rooms showed that the whole absorbing power was accurately proportional to the areas of the walls; how accurately is abundantly evidenced by the agreement of the two coefficients, .0254 and .0255, deduced on the supposition that the viscosity of the air was negligible. To express it more precisely, had the viscosity of the air been sufficient to produce one-fiftieth part of the absorption in the attendant's room, these two coefficients would have differed from each other by four per cent, an easily measurable amount. It is safe to conclude that in rooms as bare and nonabsorbent as these the viscosity of the air is inconsiderable, and that in a room filled with an audience it is certainly wholly negligible. Rooms more suitable for the demonstration of this point than these two rooms in the Boston Public Library could hardly be designed, and access to them was good fortune in settling so directly and conclusively this fundamental question.

The experiments to determine the absorbing power of plastered walls show it to be variable. If the plaster is applied directly to tile or brick the absorbing power of the resulting solid wall is uniformly .025. But if the plaster is applied to lath held out from the solid wall by studding, the absorbing power is not nearly so constant, varying in different rooms. The investigation of this has not been carried far enough to show with absolute certainty the cause, although it probably arises from the different thickness in which the plaster is applied. For the examination of this point two modes of procedure are possible, — experimenting in a large number of rooms, or experimenting in one room and replastering in many different ways. The objection to the first method, which appears the more available, is that it is almost impossible to get accurate information in regard to the nature of a wall unless one has complete control of the construction. However, there are probably interesting variations that cannot be found in use, but that, if tried, would be fruitful in suggestions for future construction. The second method — experimenting in one room, plastering and replastering it with systematic variations and careful analysis of the construction in each case — would be the most instructive, but the expense of such procedure is, for the time being at least, prohibitive. Among the interesting possibilities, of which it can only be said that the experiments so far point that way, is that with time the plastered walls improve in absorbing power; how rapidly has not been shown. This change can be due, of course, only to some real change in the nature of the wall, and the most probable change would be its gradual drying out. Experiments in four rooms with plaster on wood lath gave as the average absorbing power per square meter .034 of a unit. Experiments in eight rooms with plaster on wire lath gave as the average coefficient of absorption .033. In both cases the variation among the different rooms was such that the figure in the third decimal place may be greater or less by three, possibly, though not probably, by more. The fact that a considerable part of the wall-surface of several of the rooms was of uncertain construction is partly responsible for this uncertainty in regard to the coefficient. For the sake of easy reference and comparison these results are tabulated, the unit being the absorbing power of a square meter of open-window area.

ABSORBING POWER OF WALL-SURFACES

Open window	1.000
Wood-sheathing (hard pine)	.061
Plaster on wood lath	.034
Plaster on wire lath	.033
Glass, single thickness	.027
Plaster on tile	.025
Brick set in Portland cement	.025

Next in interest to the absorbing power of wall-surfaces is that of an audience. During the summer of 1897, at the close of a lecture in the Fogg Art Museum, the duration of the residual sound was determined before and immediately after the audience left. The patience of the audience and the silence preserved left nothing to be desired in this direction, but a slight rain falling on the roof seriously interfered with the observations. Nevertheless, the result, .37 per person, is worthy of record. The experiment was tried again in the summer of 1899, on a much more elaborate scale and under the most favorable conditions, in the large lecture-room of the Jefferson Physical Laboratory. In order to get as much data and from as independent sources as possible, three chronographs were electrically connected with each other and with the electro-pneumatic valve controlling the air supply of the organ pipe. One chronograph was on the lecture-table, and the others were on opposite sides in the rear of the hall. The one on the table was in charge of the writer, who also controlled the key turning on and off the current at the four instruments. The two other chronographs were in charge of other observers, provision being thus made for three independent determinations. After a test had been made of the absorbing power of the whole audience — 157 women and 135 men, sufficient to crowd the lecture-room — one-half, by request, passed out, 63 women and 79 men remaining, and observations were again made. On the following night the lecture was repeated and observations were again taken, there being present 95 women and 73 men. There were thus six independent determinations on three different audiences, and by three observers. In the following table the first column of figures gives the total absorbing power of the audience present; the second gives the absorbing power per person; the initials indicate the observer.

	Observer	Total Absorbing Power	Absorbing Power per Person
First night, whole audience........	W. C. S.	123.0	.42
" " " "	G. LeC.	113.0	.39
" " half "	W. C. S.	58.3	.41
" " " "	G. LeC.	58.3	.41
Second " whole "	W. C. S.	66.2	.40
" " " "	E. D. D.	64.6	.39
			.40 (3)

In view of the difficulties of the experiment the consistency of the determination is gratifying. The average result of the six determinations is probably correct within two per cent.

It is to be noted, however, that this value, .40, is the difference between the absorbing power of the person and the absorbing power of the settee and floor which, when the audience left the room, took its place as an absorbent. It is evident that the experiments determined the difference between the two, while in subsequent calculations we shall be concerned with the absolute absorbing power of the audience. To determine this, on a following night all the settees were carried out of the room, observations being taken before and after the change. From the data thus obtained the absorbing power of each settee accommodating five persons was found to be .039, or for a single seat .0077. Of necessity the floor still remained, but from a knowledge of its construction the absorbing power of as much of the floor as is covered by one person was calculated to be .030. Adding these together we get as the absorbing power of an audience, seated with moderate compactness, .44 per person.

In some subsequent work it will be necessary to know the absorbing power of an audience, not per person, but per square meter, the audience being regarded broadly as one of the bounding surfaces of the room. As each person occupied on an average .46 of a square meter of floor area, it is evident that the absorbing power per square meter was .96 of a unit.

Under certain circumstances the audience will not be compactly seated, but will be scattered about the room and more or less isolated, for example, in a council-room, or in a private music-room, and it is evident that under these conditions the individual will expose a greater surface to the room and his absorbing power will be greater. It is a matter of the greatest ease to distinguish between men and women coming into a small room, or even between different men. In fact, early in the investigation, two months' work — over three thousand observations — had to be discarded because of failure to record the kind of clothing worn by the observer. The coefficients given in the following table are averages for three women and for seven men, and were deduced from experiments in the constant-temperature room.

ABSORBING POWER OF AN AUDIENCE

Audience per square meter............................ .96
Audience per person................................. .44
Isolated woman..................................... .54
Isolated man....................................... .48

When an audience fills the hall one is but little concerned with the nature of the chairs — acoustically, but otherwise this becomes a matter of considerable importance. The settees in the lecture-room of the Physical Laboratory, already mentioned, are of plain ash, and have solid seats, and vertical ribs in the back; they are without upholstering; and it is interesting, in order to note the agreement, to compare the absorbing power of such settees per single seat, .0077, with that of the "bent wood" chairs in the Boston Public Library, .0082, which are of similar character. In contrast may be placed the chairs and settees in the faculty-room, which have cushions of hair covered with leather on seat and back. In the same table will be entered the absorbing power of Sanders Theatre cushions, which are of hair covered with canvas and light damask, and of elastic-felt cushions — cotton covered with corduroy.

ABSORBING POWER OF SETTEES, CHAIRS, AND CUSHIONS

Plain ash settees	.039
" " " per single seat	.0077
" " chairs "bent wood"	.0082
Upholstered settees, hair and leather	1.10
" " per single seat	.28
" chairs similar in style	.30
Hair cushions per seat	.21
Elastic-felt cushions per seat	.20

A case has arisen even in the present paper where it is necessary to know the absorbing power of paintings on canvas, and the question may not infrequently arise as to how much service is secured — or injury incurred — acoustically by their use in particular rooms. The oil paintings in the faculty-room, 19 in number, with a total area, 19.9 square meters, gave opportunity for the determination of the desired coefficient; but a question arises in regard to the method of reckoning the area. Thus, different coefficients are obtained according as one measures the canvas only, or includes the frames. The latter method, on the whole, seems best, although most of the absorption is probably by the canvas.

The coefficient for house plants, which may be of passing, and possibly practical, interest, was even harder to express. A greenhouse, 140 cubic meters in volume, and in which plants occupied about one-quarter of the space, showed an absorbing power greater than that due to the walls and floor by 4 units, or .11 per cubic meter of plants. It would be of greater value to determine the absorbing power of such plants as are used, often very extensively, in decorating on festival occasions, but no opportunity has yet presented itself.

Among the cloths used in decorations, cheesecloth and cretonne may be taken as types. The first is an American gauze, 48 grams to the square meter. The second is an ordinary cotton-print cloth, 182 grams per square meter. Shelia, an extra quality of chenille, is a regular curtain material used only in permanent decorations.

Linoleum and cork are commercial products, the first used as floor covering and the second in walls. Both were tested lying loosely on the floor; cemented in place, their values would probably be different.

The carpet rug is a heavy pile carpet about .8 centimeter thick.

In the following table the values are per square meter, except in the case of plants, where the coefficient is per cubic meter:

MISCELLANEOUS

Oil paintings, inclusive of frames	.28
House plants	.11
Carpet rugs	.20
Oriental rugs, extra heavy	.29
Cheesecloth	.019
Cretonne cloth	.15
Shelia curtains	.23
Hairfelt, 2.5 cm. thick, 8 cm. from wall	.78
Cork, 2.5 cm. thick, loose on floor	.16
Linoleum, loose on floor	.12

CALCULATION IN ADVANCE OF CONSTRUCTION

In the present paper it is the purpose to show the application of the preceding analysis and data, taking as an example the design of the new Boston Music Hall[1] now under construction, Messrs. McKim, Mead & White, architects.

In the introductory paper the general problem of architectural acoustics was shown to be a fairly complicated one, and to involve in its solution considerations of loudness, of interference, of resonance, and of reverberation. All these points received consideration while the Hall was being designed, but it is proposed to discuss

[1] Boston Symphony Hall.

here only the case of reverberation. In this respect a music hall is peculiarly interesting. In a theatre for dramatic performances, where the music is of entirely subordinate importance, it is desirable to reduce the reverberation to the lowest possible value in all ways not inimical to loudness; but in a music hall, concert room, or opera house, this is decidedly not the case. To reduce the reverberation in a hall to a minimum, or to make the conditions such that it is very great, may, in certain cases, present practical difficulties to the architect — theoretically it presents none. To adjust, in original design, the reverberation of a hall to a particular and approved value requires a study of conditions, of materials, and of arrangement, for which it has been the object of the preceding papers to prepare.

It is not at all difficult to show *a priori* that in a hall for orchestral music the reverberation should neither be very great, nor, on the other hand, extremely small. However, in this matter it was not necessary to rely on theoretical considerations. Mr. Gericke, the conductor of the Boston Symphony Orchestra, made the statement that an orchestra, meaning by this a symphony orchestra, is never heard to the best advantage in a theatre, that the sound seems oppressed, and that a certain amount of reverberation is necessary. An examination of all the available plans of the halls cited as more or less satisfactory models, in the preliminary discussion of the plans for the new hall, showed that they were such as to give greater reverberation than the ordinary theatre style of construction. While several plans were thus cursorily examined the real discussion was based on only two buildings — the present Boston Music Hall and the Leipzig Gewandhaus; one was familiar to all and immediately accessible, the other familiar to a number of those in consultation, and its plans in great detail were to be found in *Das neue Gewandhaus in Leipzig, von Paul Gropius und H. Schmieden*. It should, perhaps, be immediately added that neither hall served as a model architecturally, but that both were used rather as definitions and starting points on the acoustical side of the discussion. The old Music Hall was not a desirable model in every respect, even acoustically, and the Leipzig Gewandhaus, having a seating capacity about that of Sanders Theatre, 1500, was so small as to be debarred from serving directly, for this if for no other reason.

The history of the new hall is about as follows: A number of years ago, when the subject was first agitated, Mr. McKim prepared plans and a model along classical lines of a most attractive auditorium, and afterwards, at Mr. Higginson's instance, visited Europe for the purpose of consulting with musical and scientific authorities in France and Germany. But the Greek Theatre as a music hall was an untried experiment, and because untried was regarded as of uncertain merits for the purpose by the conductors consulted by Mr. Higginson and Mr. McKim. It was, therefore, abandoned. Ten years later, when the project was again revived, the conventional rectangular form was adopted, and the intention of the building committee was to follow the general proportions and arrangement of the Leipzig Gewandhaus, so enlarged as to increase its seating capacity about seventy per cent; thus making it a little more than equal to the old hall. At this stage calculation was first applied.

The often-repeated statement that a copy of an auditorium does not necessarily possess the same acoustical qualities is not justified, and invests the subject with an unwarranted mysticism. The fact is that exact copies have rarely been made, and can hardly be expected. The constant changes and improvements in the materials used for interior construction in the line of better fireproofing — wire lath or the application of the plaster directly to tile walls — have led to the taking of liberties in what were perhaps regarded as nonessentials; this has resulted, as shown by the tables, in a changed absorbing power of the walls. Our increasing demands in regard to heat and ventilation, the restriction on the dimensions enforced by location, the changes in size imposed by the demands for seating capacity, have prevented, in different degrees, copies from being copies, and models from successfully serving as models. So different have been the results under what was thought to be safe guidance — but a guidance imperfectly followed — that the belief has become current that the whole subject is beyond control. Had the new Music Hall been enlarged from the Leipzig Gewandhaus to increase the seating capacity seventy per cent, which, proportions being preserved, would have doubled the volume, and then built, as it is being built, according to the most modern methods of fireproof construction, the result, unfortunately, would have been to confirm the belief. No mistake is more easy to make than that of copying an auditorium — but in different materials or on a different scale — in the expectation that the result will be the same. Every departure must be compensated by some other — a change in material by a change in the size or distribution of the audience, or perhaps by a partly compensating change in the material used in some other part of the hall — a change in size by a change in the proportions or shape. For moderate departures from the model such compensation can be made, and the model will serve well as a guide to a first approximation. When the departure is great the approved auditorium, unless discriminatingly used, is liable to be a treacherous guide. In this case the departure was necessarily great.

The comparison of halls should be based on the duration of the residual sound after the cessation of a source that has produced over the hall some standard average intensity of sound, — say one million times the minimum audible intensity, 1,000,000 i'. The means for this calculation was furnished in the fifth paper. The values of V and a for the three halls under comparison are shown on the next page.

The length given for the Leipzig Gewandhaus, 38 meters, is measured from the organ front to the architecturally principal wall in the rear. On the floor and by boxes in the balconies the seats extend 3 meters farther back, making the whole length of the hall, exclusive of the organ niche, 41 meters. This increases the volume of the hall about 200 cubic meters, making the total volume 11,400 cubic meters.

The height given for the new Boston Music Hall, 17.9, is the average height from the sloping floor. The length is measured on the floor of the main part of the hall; above the second gallery it extends back 2.74 meters, giving an additional volume of 580 cubic meters. The stage, instead of being out in the room, is in a contracted recess having a depth of 7.9 meters, a breadth, front and back, of 18.3 and 13.6, respectively, and a height, front and back, of 13.4 and 10.6, respectively, with a volume of 1,500 cubic meters. The height of the stage recess is determined by the absolute re-

DIMENSIONS OF THE THREE HALLS IN METERS [1]

	Leipzig Gewandhaus	Boston Music Hall, Old	Boston Music Hall, New
Length	(38)	39.2	(39.5)
Breadth	19	23.5	22.8
Height	15.5	20.0	17.9
Volume	(11,200)	18,400	(16,200)

quirements of the large organ to be built by Mr. George S. Hutchings. This organ will extend across the whole breadth of the stage. The total volume of the new Boston Music Hall is, therefore, 18,300 cubic meters.

In the following table of materials in the three halls no distinction is made between plaster on wire lath and plaster on wood lath, the experiments recorded in the preceding paper having shown no certain difference in absorbing power. The areas of wall-surface are expressed in square meters. The number of persons in the audience is reckoned from the number of seats, no account being taken of standing room.

[1] DIMENSIONS OF THE THREE HALLS IN FEET

	Leipzig Gewandhaus	Boston Music Hall, Old	Boston Music Hall, New
Length	(124)	129	(130)
Breadth	62	77	75
Height	52	66	59
Volume	(400,000)	656,000	(575,000)

The length given for the Leipzig Gewandhaus, 124 feet, is measured from the organ front to the architecturally principal wall in the rear. On the floor and by boxes in the balconies the seats extend 10 feet farther back, making the total length of the hall, exclusive of the organ niche, 134 feet. This increases the volume 7,000 cubic feet, making the total volume 407,000 cubic feet.

The height given for the new hall, 59 feet, is the average height from the sloping floor. The length is measured on the floor of the main part of the hall; above the second gallery it extends back 9 feet, giving an additional volume of 20,000 cubic feet. The stage, instead of being out in the room, is in a contracted recess, having a depth of 26 feet, a breadth, front and back, of 60 feet and 45 feet, respectively, and a height, front and back, of 44 feet and 35 feet, respectively, with a volume of 54,000 cubic feet. The total volume of the new Music Hall is, therefore, 649,000 cubic feet.

ABSORBING MATERIAL

	Leipzig Gewandhaus	Boston Music Hall, Old	Boston Music Hall, New
Plaster on lath	2,206	3,030	1,040
Plaster on tile	0	0	1,830
Glass	17	55	22
Wood	235	771	625
Drapery	80	4	0
Audience:			
on floor	990	1,251	1,466
in 1st balcony	494	680	606
in 2d balcony	33	460	507
Total audience	1,517	2,391	2,579
Orchestra	80	80	80

The drapery in the Leipzig Gewandhaus will be rated as shelia, and in the old Music Hall as cretonne, to which it approximates in each case. It is an almost needless refinement to rate differently the orchestra and the audience merely because the members of the orchestra sit more or less clear of each other, but for the sake of a certain formal completeness it will be done. For the above materials the coefficients, taken from the preceding paper, are as follows:

COEFFICIENTS OF ABSORPTION

Plaster on lath	.033
Plaster on tile	.025
Glass	.027
Wood	.061
Drapery { shelia	.23
cretonne	.15
Audience per person	.44
Orchestra per man	.48

In the table (p. 67) is entered the total absorbing power contributed by each of these elements. As this is the first example of such calculation all the elements will be shown, although it will then be immediately evident that some are of wholly negligible magnitude.

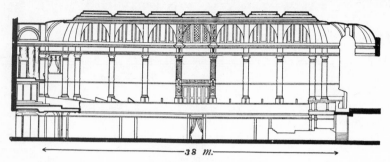

FIG. 20.   The Leipzig Gewandhaus.

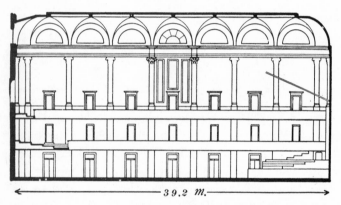

FIG. 21.   The Old Boston Music Hall.

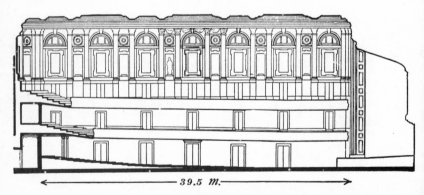

FIG. 22.   The New Boston Music Hall.

ABSORBING POWER

	Leipzig Gewandhaus	Boston Music Hall, Old	Boston Music Hall, New
Plaster on lath	73	100	34
Plaster on tile	0	0	46
Glass	0.4	1.5	0.6
Wood	14	47	38
Drapery	18	0.6	0
Audience	667	1,052	1,135
Orchestra	38	38	38
Total = a	810	1,239	1,292

V and a being determined for each of the three halls, the duration, T, of the residual sound after standard initial intensity can be calculated.

The results, in seconds, are as follows:

Leipzig Gewandhaus 2.30
Old Boston Music Hall 2.44
New Boston Music Hall 2.31

In other words, the new hall, although having a seating capacity for over a thousand more than the Gewandhaus and nearly two hundred more than the old hall, will have a reverberation between the two, and nearer that of the Gewandhaus than that of the old hall.

It is interesting to contrast this with the result that would have been obtained had the plan been followed of reproducing on an enlarged scale the Gewandhaus. Assuming perfect reproduction of all proportions with like materials, the volume would have been 25,300 cubic meters, and the absorbing power 1,370, resulting in the value, $T = 3.02$. This would have differed from the chosen result by an amount that would have been very noticeable.

The new Boston Music Hall is, therefore, not a copy of the Gewandhaus, but the desired results have been attained in a very different way.

A few general considerations, not directly connected with reverberation, may be of interest. The three halls are of nearly the same length on the floor; but in the old hall and in the Gewandhaus the platform for the orchestra is out in the hall, and the galleries extend along both sides of it; while in the new hall the orchestra is not out in the main body of the room, and for this reason is slightly farther from the rear of the hall; but this is more than compensated for in respect to loudness by the orchestra being in a somewhat contracted stage recess, from the side walls of which the reflection is better because they are nearer and not occupied by an audience. Also it may be noted that the new hall is not so high as the old and is not so broad.

Thus is opened up the question of loudness, and this has been solved to a first approximation for the case of sustained tones. But as the series of papers now concluded is devoted to the question of reverberation, this new problem must be reserved for a subsequent discussion.

2

ARCHITECTURAL ACOUSTICS[1]

INTRODUCTION

THE problem of architectural acoustics requires for its complete solution two distinct lines of investigation, one to determine quantitatively the physical conditions on which loudness, reverberation, resonance, and the allied phenomena depend, the other to determine the intensity which each of these should have, what conditions are best for the distinct audition of speech, and what effects are best for music in its various forms. One is a purely physical investigation, and its conclusions should be based and should be disputed only on scientific grounds; the other is a matter of judgment and taste, and its conclusions are weighty in proportion to the weight and unanimity of the authority in which they find their source. For this reason, these papers are in two series. The articles which appeared six years ago began the first, and the paper immediately following is the beginning of the second.

Of the first series of papers, which have to do with the purely physical side of the problem, only one paper has as yet been published. This contained a discussion of reverberation, complete as far as one note is concerned. There is on hand considerable material for a paper extending this discussion to cover the whole range of the musical scale, and therefore furnishing a basis for the discussion of what has sometimes been called the musical quality of an auditorium. There has also been collected a certain amount of data in regard to loudness, resonance, interference, echoes, irregularities of air currents and temperature, and the transmission of sound through walls and partitions, — all of which will appear as soon as a complete presentation is possible in each case. Each problem has been taken up as it has been brought to the writer's attention by an architect in consultation either over plans or in regard to a completed building. This method is slow, but it has the advantage of

[1] Proceedings of the American Academy of Arts and Sciences, vol. xlii, no. 2, June, 1906.

making the work practical, and may be relied on to prevent the magnification to undue importance of scientifically interesting but practically subordinate points. On the other hand, there is the danger that it may lead to a fragmentary presentation. An effort has been made to guard against this, and the effort for completeness is the reason for delay in the appearance of some of the papers. Sufficient progress has been made, however, to justify the assertion that the physical side of the problem is solvable, and that it should be possible ultimately to calculate in advance of construction all the acoustical qualities of an auditorium.

Thus far it is a legitimate problem in physics, and as such a reasonable one for the writer to undertake.

The second part of the problem, now being started, the question as to what constitutes good and what constitutes poor acoustics, what effects are desirable in an auditorium designed for speaking, and even more especially in one designed for music, is not a question in physics. It is therefore not one for which the writer is especially qualified, and would not be undertaken here were it not in the first place absolutely necessary in order to give effect to the rest of the work, and in the second place were it not the plan rather to gather and give expression to the judgment of others acknowledged as qualified to speak, than to give expression to the taste and judgment of one. It is thus the purpose to seek expert judgment in regard to acoustical effects, and if possible to present the results in a form available to architects. This will be slow and difficult work, and it is not at all certain that it will be possible to arrive, even ultimately, at a finished product. It is worth undertaking, however, if the job as a whole is worth undertaking, for without it the physical side of the investigation will lose much of its practical value. Thus it is of little value to be able to calculate in advance of construction and express in numerical measure the acoustical quality which any planned auditorium will have, unless one knows also in numerical measure the acoustical quality which is desired. On the other hand, if the owner and the architect can agree on the desired result, and if this is within the limits of possibility considering all the demands on the auditorium, of utility, architecture, and engineering, this result can be secured with certainty, — at least there need be no uncertainty as to whether it will or will not be attained in the completed building.

The papers following this introduction will be: *The Accuracy of Musical Taste in regard to Architectural Acoustics*, and *Variation in Reverberation with Variation in Pitch*.

THE ACCURACY OF MUSICAL TASTE IN REGARD TO ARCHITECTURAL ACOUSTICS

PIANO MUSIC

THE experiments described in this paper were undertaken in order to determine the reverberation best suited to piano music in a music room of moderate size, but were so conducted as to give a measure of the accuracy of cultivated musical taste. The latter point is obviously fundamental to the whole investigation, for unless musical taste is precise, the problem, at least as far as it concerns the design of the auditorium for musical purposes, is indeterminate.

The first observations in regard to the precision of musical taste were obtained during the planning of the Boston Symphony Hall, Messrs. McKim, Mead, and White, Architects. Mr. Higginson, Mr. Gericke, the conductor of the orchestra, and others connected with the Building Committee expressed opinions in regard to a number of auditoriums. These buildings included the old Boston Music Hall, at that time the home of the orchestra, and the places visited by the orchestra in its winter trips, Sanders Theatre in Cambridge, Carnegie Hall in New York, the Academy of Music in Philadelphia, and the Music Hall in Baltimore, and in addition to these the Leipzig Gewandhaus. By invitation of Mr. Higginson, the writer accompanied the orchestra on one of its trips, made measurements of all the halls, and calculated their reverberation. The dimensions and the material of the Gewandhaus had been published, and from these data its reverberation also was calculated. The results of these measurements and calculations showed that the opinions expressed in regard to the several halls were entirely consistent with the physical facts. That is to say, the reverberation in those halls in which it was declared too great was in point of physical measurement greater than in halls in which it was pronounced too small. This consistency gave encouragement in the hope that the physical problem was real, and the end to be attained definite.

Much more elaborate data on the accuracy of musical taste were obtained four years later, 1902, in connection with the new building of the New England Conservatory of Music, Messrs. Wheelwright and Haven, Architects. The new building consists of a large auditorium surrounded on three sides by smaller rooms, which on the second and third floors are used for purposes of instruction. These smaller rooms, when first occupied, and used in an unfurnished or partially furnished condition, were found unsuitable acoustically, and the writer was consulted by Mr. Haven in regard to their final adjustment. In order to learn the acoustical condition which would accurately meet the requirements of those who were to use the rooms, an experiment was undertaken in which a number of rooms, chosen as typical, were varied rapidly in respect to reverberation by means of temporarily introduced absorbing material. Approval or disapproval of the acoustical quality of each room at each stage was expressed by a committee chosen by the Director of the Conservatory. At the close of these tests, the reverberation in the rooms was measured by the writer in an entirely independent manner as described in the paper on Reverberation (1900). The judges were Mr. George W. Chadwick, Director of the Conservatory, and Signor Oresti Bimboni, Mr. William H. Dunham, Mr. George W. Proctor, and Mr. William L. Whitney, of the Faculty. The writer suggested and arranged the experiment and subsequently reduced the results to numerical measure, but expressed no opinion in regard to the quality of the rooms.

The merits of each room in its varied conditions were judged solely by listening to piano music by Mr. Proctor. The character of the musical compositions on which the judgment was based is a matter of interest in this connection, but this fact was not appreciated at the time and no record of the selections was made. It is only possible to say that several short fragments, varied in nature, were tried in each room.

As will be evident from the descriptions given below, the rooms were so differently furnished that no inference as to the reverberation could be drawn from appearances, and it is certain that the opinions were based solely on the quality of the room as heard in the piano music.

The five rooms chosen as typical were on the second floor of the building. The rooms were four meters high. Their volumes varied from 74 to 210 cubic meters. The walls and ceilings were finished in plaster on wire lath, and were neither papered nor painted. There was a piano in each room; in room 5 there were two. The amount of other furniture in the rooms varied greatly:

In room 1 there was a bare floor, and no furniture except the piano and piano stool.

Room 2 had rugs on the floor, chairs, a sofa with pillows, table, music racks, and a lamp.

Room 3 had a carpet, chairs, bookcases, and a large number of books, which, overflowing the bookcases, were stacked along the walls.

Room 4 had no carpet, but there were chairs and a small table.

Room 5 had a carpet, chairs, and shelia curtains.

Thus the rooms varied from an almost unfurnished to a reasonably furnished condition. In all cases the reverberation was too great.

The experiment was begun in room 1. There were, at the time, besides the writer, five gentlemen in the room, the absorbing effect of whose clothing, though small, nevertheless should be taken into account in an accurate calculation of the reverberation. Thirteen cushions from the seats in Sanders Theatre, whose absorbing power for sound had been determined in an earlier investigation, were brought into the room. Under these conditions the unanimous opinion was that the room, as tested by the piano, was lifeless. Two cushions were then removed from the room with a perceptible change for the better in the piano music. Three more cushions were removed, and the effect was much better. Two more were then taken out, leaving six cushions in the room, and the result met unanimous approval. It was suggested that two more be removed. This being done the reverberation was found to be too great. The agreement was then reached that the conditions produced by the presence of six cushions were the most nearly satisfactory.

The experiment was then continued in Mr. Dunham's room, number 2. Six gentlemen were present. Seven cushions were brought into the room. The music showed an insufficient reverberation. Two of the cushions were then taken out. The change was regarded as a distinct improvement, and the room was satisfactory.

In Mr. Whitney's room, number 3, twelve cushions, with which it was thought to overload the room, were found insufficient even with the presence in this case of seven gentlemen. Three more cushions were brought in and the result declared satisfactory.

In the fourth room, five, eight, and ten cushions were tried before the conditions were regarded as satisfactory.

In Mr. Proctor's room, number 5, it was evident that the ten cushions which had been brought into the room had overloaded it. Two were removed, and afterwards three more, leaving only five, before a satisfactory condition was reached.

This completed the direct experiment with the piano.

The bringing into a room of any absorbing material, such as these cushions, affects its acoustical properties in several respects, but principally in respect to its reverberation. The prolongation of sound in a room after the cessation of its source may be regarded either as a case of stored energy which is gradually suffering loss by transmission through and absorption by the walls and contained material, or it may be regarded as a process of rapid reflection from wall to wall with loss at each reflection. In either case it is called reverberation. It is sometimes called, mistakenly as has been explained, resonance. The reverberation may be expressed by the duration of audibility of the residual sound after the cessation of a source so adjusted as to produce an average of sound of some standard intensity over the whole room. The direct determination of this, under the varied conditions of this experiment, was impracticable, but, by measuring the duration of audibility of the residual sound after the cessation of a measured organ pipe in each room without any cushions, and knowing the coefficient of absorption of the cushions, it was possible to calculate accurately the reverberation at each stage in the test. It was impossible to make these measurements immediately after the above experiments, because, although the day was an especially quiet one, the noises from the street and railway traffic were seriously disturbing. Late the following night the conditions were more favorable, and a series of fairly good observations was obtained in each room. The cushions had been removed, so that the measurements were made on the rooms in their original condition, furnished as above described. The apparatus and method employed are described in full in a series of articles in the Engineering Record [1] and American Architect for 1900. The results are given in the accompanying table.

Room Number	Volume	Absorbing Power of Room	Gentlemen Present	Absorbing Power of Clothing	Number of Meters of Cushions	Absorbing Power of Cushions	Total Absorbing Power	Reverberation in Seconds	Remarks
1	74	5.0	0	0	0	0	5.0	2.43	Reverberation too great.
		"	5	2.4	0	0	7.4	1.64	Reverberation too great.
		"	"	"	13	12.8	20.2	.60	Reverberation too little.
		"	"	"	11	10.1	17.5	.70	Better.
		"	"	"	8	7.3	14.7	.83	Better.
		"	"	"	6	5.5	12.9	.95	Condition approved.
		"	"	"	4	3.6	11.0	1.22	Reverberation too great.
2	91	6.3	0	0	0	0	6.3	2.39	Reverberation too great.
		"	6	2.9	0	0	9.2	1.95	Reverberation too great.
		"	"	"	7	6.4	15.6	.95	Reverberation too little.
		"	"	"	5	4.6	13.8	1.10	Condition approved.
3	210	14.0	0	0	0	0	14.0	2.46	Reverberation too great.
		"	7	3.4	0	0	17.4	2.00	Reverberation too great.
		"	"	"	12	11.0	28.4	1.21	Better.
		"	"	"	15	13.7	31.1	1.10	Condition approved.
4	133	8.3	0	0	0	0	8.3	2.65	Reverberation too great.
		"	7	3.4	0	0	11.7	1.87	Reverberation too great.
		"	"	"	6	5.5	17.2	1.26	Better.
		"	"	"	10	9.1	20.8	1.09	Condition approved.
5	96	7.0	0	0	0	0	7.0	2.24	Reverberation too great.
		"	4	1.9	0	0	8.9	1.76	Reverberation too great.
		"	"	"	10	9.1	18.0	.87	Reverberation too little.
		"	"	"	8	7.3	16.2	.98	Better.
		"	"	"	5	4.6	13.5	1.16	Condition approved.

[1] The article in the Engineering Record is identical with the paper in the American Architect for 1900, reprinted in this volume as Part I.

The table is a record of the first of what, it is hoped, will be a series of such experiments extending to rooms of much larger dimensions and to other kinds of music. It may well be, in fact it is highly probable, that very much larger rooms would necessitate a different amount of reverberation, as also may other types of musical instruments or the voice. As an example of such investigations, as well as evidence of their need, it is here given in full. The following additional explanations may be made. The variation in volume of the rooms is only threefold, corresponding only to such music rooms as may be found in private houses. Over this range a perceptible variation in the required reverberation should not be expected. The third column in the table includes in the absorbing power of the room (ceiling, walls, furniture, etc.) the absorbing powers of the clothes of the writer, who was present not merely at all tests, but in the measurement of the reverberation the following night. From the next two columns, therefore, the writer and the effects of his clothing are omitted. The remarks in the last column are reduced to the form "reverberation too great," "too little," or "approved." The remarks at the time were not in this form, however. The room was pronounced "too resonant," "too much echo," "harsh," or "dull," "lifeless," "overloaded," expressions to which the forms adopted are equivalent.

If from the larger table the reverberation in each room, in its most approved condition, is separately tabulated, the following is obtained:

Rooms	Reverberation
1	.95
2	1.10
3	1.10
4	1.09
5	1.16

1.08 mean

The final result obtained, that the reverberation in a music room in order to secure the best effect with a piano should be 1.08, or in round numbers 1.1, is in itself of considerable practical value; but the five determinations, by their mutual agreement, give a numerical measure to the accuracy of musical taste which is of great interest. Thus the maximum departure from the mean is .13 seconds, and the average departure is .05 seconds. Five is rather a small number of observations on which to apply the theory of probabilities, but, assuming that it justifies such reasoning, the probable error is .02 seconds, — surprisingly small.

A close inspection of the large table will bring out an interesting fact. The room in which the approved condition differed most from the mean was the first. In this room, and in this room only, was it suggested by the gentlemen present that the experiment should be carried further. This was done by removing two more cushions. The reverberation was then 1.22 seconds, and this was decided to be too much. The point to be observed is that 1.22 is further above the mean, 1.08, than .95 is below. Moreover, if one looks over the list in each room it will be seen that in every case the reverberation corresponding to the chosen condition came nearer to the mean than that of any other condition tried.

It is conceivable that had the rooms been alike in all respects and required the same amount of cushions to accomplish the same results, the experiment in one room might have prejudiced the experiment in the next. But the rooms being different in size and furnished so differently, an impression formed in one room as to the number of cushions necessary could only be misleading if depended on in the next. Thus the several rooms required 6, 5, 15, 10, and 5 cushions. It is further to be observed that in three of the rooms the final condition was reached in working from an overloaded condition, and in the other two rooms from the opposite condition, — in the one case by taking cushions out, and in the other by bringing them in.

Before beginning the experiment no explanation was made of its nature, and no discussion was held as to the advantages and disadvantages of reverberation. The gentlemen present were asked to express their approval or disapproval of the room at each stage of the experiment, and the final decision seemed to be reached with perfectly free unanimity.

This surprising accuracy of musical taste is perhaps the explanation of the rarity with which it is entirely satisfied, particularly when the architectural designs are left to chance in this respect.

VARIATION IN REVERBERATION WITH VARIATION IN PITCH

SIX years ago there was published in the Engineering Record and the American Architect a series of papers on architectural acoustics intended as a beginning in the general subject. The particular phase of the subject under consideration was reverberation,—the continuation of sound in a room after the source has ceased. It was there shown to depend on two things,—the volume of the room, and the absorbing character of the walls and of the material with which the room is filled. It was also mentioned that the reverberation depends in special cases on the shape of the room, but these special cases were not considered. The present paper also will not take up these special cases, but postpone their consideration, although a good deal of material along this line has now been collected. It is the object here to continue the earlier work rather narrowly along the original lines. The subject was then investigated solely with reference to sounds of one pitch, C_4 512 vibrations per second. It is the intention here to extend this over nearly the whole range of the musical scale, from C_1 64 to C_7 4096.

It can be shown readily that the various materials of which the walls of a room are constructed and the materials with which it is filled do not have the same absorbing power for all sounds regardless of pitch. Under such circumstances the previously published work with C_4 512 must be regarded as an illustration, as a part of a much larger problem, — the most interesting part, it is true, because near the middle of the scale, but after all only a part. Thus a room may have great reverberation for sounds of low pitch and very little for sounds of high pitch, or exactly the reverse; or a room may have comparatively great reverberation for sounds both of high and of low pitch and very little for sounds near the middle of the scale. In other words, it is not putting it too strongly to say that a room may have very different quality in different registers, as different as does a musical instrument; or, if the room is to be used for speaking purposes, it may have different degrees of excellence or defect for a whisper and for the full rounded tones of the voice, different for a woman's voice and for a man's — facts more or less well recognized. Not to leave this as a vague generalization the following cases may be cited. Recently, in discussing the acoustics of the proposed cathedral of southern California in Los Angeles with Mr. Maginnis, its architect, and the writer, Bishop Conaty touched on this point very clearly. After discussing the general subject with more than the usual insight and experience, possibly in part because Catholic churches and cathedrals have great reverberation, he added that he found it difficult to avoid pitching his voice to that note which the auditorium most prolongs notwithstanding the fact that he found this the worst pitch on which to speak. This brings out, perhaps more impressively because from practical experience instead of from theoretical considerations, the two truths that auditoriums have very different reverberation for different pitches, and that excessive reverberation is a great hindrance to clearness of enunciation. Another incident may also serve, that of a church near Boston, in regard to which the writer has just been consulted. The present pastor, in describing the nature of its acoustical defects, stated that different speakers had different degrees of difficulty in making themselves heard; that he had no difficulty, having a rather high pitched voice; but that the candidate before him, with a louder but much lower voice, failed of the appointment because unable to make himself heard. Practical experience of the difference in reverberation with variation of pitch is not unusual, but the above cases are rather striking examples. Corresponding effects are not infrequently observed in halls devoted to music. Its observation here, however, is marked in the rather complicated general effect. The full discussion of this belongs to another series of papers, in which will be taken up the subject of the acoustical effects or conditions that are desirable for music and for speech. While this phase of the subject will not be discussed here at length, a little consideration of the data to be presented will show how pronounced these effects may be and how important in the general subject of architectural acoustics.

In order to show the full significance of this extension of the investigation in regard to reverberation, it is necessary to point out some features which in earlier papers were not especially emphasized. Primarily the investigation is concerned with the subject of reverberation, that is to say, with the subject of the continuation of a sound in a room after the source has ceased. The immediate effect of reverberation is that each note, if it be music, each syllable or part of a syllable, if it be speech, continues its sound for some time, and by its prolongation overlaps the succeeding notes or syllables, harmoniously or inharmoniously in music, and in speech always towards confusion. In the case of speech it is inconceivable that this prolongation of the sound, this reverberation, should have any other effect than that of confusion and injury to the clearness of the enunciation. In music, on the other hand, reverberation, unless in excess, has a distinct and positive advantage.

Perhaps this will be made more clear, or at least more easily realized and appreciated, if we take a concrete example. Given a room comparatively empty, with hard wall-surfaces, for example plaster or tile, and having in it comparatively little furniture, the amount of reverberation for the sounds of about the middle register of the double-bass viol and for the sounds of the middle register of the violin will be very nearly though not exactly equal. If, however, we bring into the room a quantity of elastic felt cushions, sufficient, let us say, to accommodate a normal audience, the effect of these cushions, the audience being supposed absent, will be to diminish very much the reverberation both for the double-bass viol and for the violin, but will diminish them in very unequal amounts. The reverberation will now be twice as great for the double-bass as for the violin. If an audience comes into the room, filling up the seats, the reverberation will be reduced still further and in a still greater disproportion, so that with an audience entirely filling the room the reverberation for the violin will be less than one-third that for the double-bass. When one considers that a difference of five per cent in reverberation is a matter for approval or disapproval on the part of musicians of critical taste, the importance of considering these facts is obvious.

This investigation, nominally in regard to reverberation, is in reality laying the foundation for other phases of the problem. It has as one of its necessary and immediate results a determination of the coefficient of absorption of sound of various materials. These coefficients of absorption, when once known, enable one not merely to calculate the prolongation of the sound, but also to calculate the average loudness of sustained tones. Thus it was shown in one of the earlier papers, though at that time no very great stress was laid on it, that the average loudness of a sound in a room is proportional inversely to the absorbing power of the material in the room. Therefore the data which are being presented, covering the whole range of the musical scale, enable one to calculate the loudness of different notes over that range, and make it possible to show what effect the room has on the piano or the orchestra in different parts of the register.

To illustrate this by the example above cited, if the double-bass and the violin produce the same loudness in the open air, in the bare room with hard walls both would be reënforced about equally. The elastic felt brought into the room would decidedly diminish this reenforcement for both instruments. It would, however, exert a much more pronounced effect in the way of diminishing the reënforcement for the violin than for the double-bass. In fact, the balance will be so affected that it will require two violins to produce the same volume of sound as does one double-bass. The audience coming into the room will make it necessary to use three violins to a double-bass to secure the same balance as before.

Both cases cited above are only broadly illustrative. As a matter of fact the effect of the room and the effect of the audience in the room is perceptibly different at the two ends of the register of the violin and of the double-bass viol.

There is still a third effect, which must be considered to appreciate fully the practical significance of the results that are being presented. This is the effect on the quality of a sustained tone. Every musical tone is composed of a great number of partial tones, the predominating one being taken as the fundamental, and its pitch as the pitch of the sound. The other partial tones are regarded as giving quality or color to the fundamental. The musical quality of a tone depends on the relative intensities of the overtones. It has been customary, at least on the part of physicists, to regard the relative intensities of the overtones, which define the quality of the sound, as depending simply on the source from which the sound originates. Of course, primarily, this is true. Nevertheless, while the source defines the relative intensities of the issuing sounds, their actual intensities in the room depend not merely on that, but also, and to a surprising degree, on the room itself. Thus, for example, given an eight-foot organ pipe, if blown in an empty room, such as that described above, the overtones would be pronounced. If exactly the same pipe be blown with the same wind pressure in a room in which the seats have been covered with the elastic felt, the first upper partial will bear to the fundamental a ratio of intensity diminished over 40 per cent, the second upper partial a ratio to the fundamental diminished in the same per cent, the third upper partial a ratio diminished over 50 per cent, while the fourth upper partial will bear a ratio of intensity to the fundamental diminished about 60 per cent. Quality expressed numerically in this way probably does not convey a very vivid impression as to its real effect. It may signify more to say merely that the change in quality is very pronounced and noticeable, even to comparatively untrained ears. On the other hand, if one were to try the experiment with a six-inch instead of with an eight-foot organ pipe, the effect of bringing the elastic felt cushions into the room would be to increase the relative intensities of the overtones, and thus to diminish the purity of the tone.

All tones below that of a six-inch organ pipe will be purified by bringing into the room elastic felt. All tones above and including that pitch will be rendered less pure. The effect of an audience coming into a room is still different. Assuming that the audience has filled the room and so covered all the elastic felt cushions, the effect of the audience is to purify all tones up to violin C_4 512, and to have very little effect on all tones from that pitch upward. On very low tones the effect of the audience in the room is more pronounced. For example, again take C_1 64, the effect of the audience will be to diminish its first overtone about 60 per cent relative to the fundamental and its second overtone over 75 per cent.

The effect of the material used in the construction of a room, and the contained furniture, in altering the relative intensities of the fundamental and the overtones, is to improve or injure its quality according to circumstances. It may be, of course, that the tone desired is a very pure one, or it may be that what is wanted is a tone with pronounced upper partials. Take, for example, the "night horn" stop in a pipe organ. This is intended to have a very pure tone. The room in contributing to its purity would improve its quality. On the other hand, the mixture stop in a pipe organ is intended to have very pronounced overtones. In fact to this end not one but several pipes are sounded at once. The effect of the above room to emphasize the fundamental and to wipe out the overtones would be in opposition to the original design of the stop. To determine what balance is desirable must lie of course with the musicians. The only object of the present series of papers is to point out the fundamental facts, and that our conditions may be varied in order to attain any desired end. One great thing needed is that the judgment of the musical authorities should be gathered in an available form; but that is another problem, and the above bare outline is intended only to indicate the importance of extending the work to the whole range of the musical scale, — the work undertaken in the present paper.

The method pursued in these experiments is not very unlike that followed in the previous experiments with C_4 512. It differs in minor detail, but to explain these details would involve a great deal of repetition which the modifications in the method are not of sufficient importance to justify.

Broadly, the procedure consists first in the determination of the rate of emission of the sound of an organ pipe for each note to be investigated. This consists in determining the durations of audibility after the cessation of two sounds, one having four or more, but a known multiple, times the intensity of the other. From these results it is possible to determine the rate of emission by the pipes, each in terms of the minimum audibility for that particular tone. The apparatus used in this part of the experiment is shown in Fig. 1. Four small organs were fixed at a minimum distance of five meters apart. It was necessary to place them at this great distance apart because, as already pointed out, if placed near each other the four sounded together do not emit four times the sound emitted by one. This wide separation was particularly necessary for the large pipes and the low tones; a very much less separation would have served the purpose in the case of the high tones.

From the point where the four tubes leading to the small organs meet, a supply pipe ran, as shown on the drawing, to an air reservoir in the room below. This was fed from an electrically driven blower at the far end of the building. The chronograph was in another room. The experiments with this apparatus, like the experiments

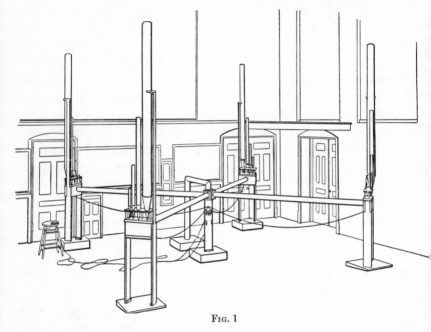

Fig. 1

heretofore recorded, were carried out at night between twelve and five o'clock.

The rate of emission of sound by the several pipes having been determined, the next work was the determination of the coefficients of absorption. The methods employed having already been sufficiently described, only results will be given.

In the very nature of the problem the most important data is the absorption coefficient of an audience, and the determination of this was the first task undertaken. By means of a lecture on one of the recent developments of physics, an audience was enveigled into attending, and at the end of the lecture requested to remain for the experiment. In this attempt the effort was made to determine the coefficients for the five octaves from C_2 128 to C_6 2048, including notes E and G in each octave. For several reasons the experiment was not a success. A threatening thunder storm made the audience a small one, and the sultriness of the atmosphere made open windows necessary, while the attempt to cover so many notes, thirteen in all, prolonged the experiment beyond the endurance of the audience. While this experiment failed, another the following summer was more successful. In the year that had elapsed the necessity of carrying the investigation further than the limits intended became evident, and now the experiment was carried from C_1 64 to C_7 4096, but including only the C notes, seven notes in all. Moreover, bearing in mind the experiences of the previous summer, it was recognized that even seven notes would come dangerously near overtaxing the patience of the audience. Inasmuch as the coefficient of absorption for C_4 512 had already been determined six years before in the investigations mentioned, the coefficient for this note was not redetermined. The experiment was therefore carried out for the lower three and the upper three notes of the seven. The audience, on the night of this experiment, was much larger than that which came the previous summer, the night was a more comfortable one, and it was possible to close the windows during the experiment. The conditions were thus fairly satisfactory. In order to get as much data as possible and in as short a time, there were nine observers stationed at different points in the room. These observers, whose kindness and skill it is a pleasure to acknowledge, had prepared themselves by previous practice for this one experiment. As in the work of six years ago, the writer's key controlled the organ pipes and started the chronograph, the writer and the other observers each had a key which was connected with the chronograph to record the cessation of audibility of the sound. The results of the experiment are shown on the lower curve in Fig. 2. This curve gives the coefficient of absorption per person. It is to be observed that one of the points falls clearly off the smooth curve drawn through the other points. The observations on which this point is based were, however, much disturbed by a street car passing not far from the building, and the departure of this observation from the curve does not indicate a real departure in the coefficient nor should it cast much doubt on the rest of the work, in view of the circumstances under which it was secured. Counteracting the perhaps bad impression which this point may give, it is a considerable satisfaction to note how accurately the point for C_4 512, deter-

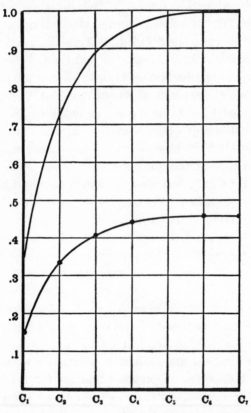

Fig. 2. The absorbing power of an audience for different notes. The lower curve represents the absorbing power of an audience per person. The upper curve represents the absorbing power of an audience per square meter as ordinarily seated. The vertical ordinates are expressed in terms of total absorption by a square meter of surface. For the upper curve the ordinates are thus the ordinary coefficients of absorption. The several notes are at octave intervals, as follows: $C_1$64, $C_2$128, C_3 (middle C) 256, $C_4$512, $C_5$1024, $C_6$2048, $C_7$4096.

mined six years before by a different set of observers, falls on the smooth curve through the remaining points. In the audience on which these observations were taken there were 77 women and

105 men. The courtesy of the audience in remaining for the experiment and the really remarkable silence which they maintained is gratefully acknowledged.

The curve above discussed is that for the average person in an audience. An interesting form in which to throw the results is to regard the audience as one side of a room. We may then look at it as an extended absorbing surface, and determine the coefficient per square meter. Worked out on this basis the absorption coefficient is indicated in the higher curve. It is merely the lower curve multiplied by a number which expresses the average number of people per square meter. It is interesting to note that the coefficient of absorption is about the same from C_4 512 up, indicating over that range nearly complete absorption. Below that point there is a very great falling off, down to C_1 64. The curve is such as to permit of an extrapolation indicative of even less absorption and consequently greater reverberation for the still lower notes. Without entering into an elaborate discussion of this curve, two points may be noted as particularly interesting. The first is the nearly complete absorption for the higher notes, a result which at first sight seems a little inconsistent with the results which will be shown later on in connection with the absorption by felt. The inconsistency, however, is only apparent. The greater absorption shown by an audience than that shown by thick felt arises from the fact that the surface of the audience is irregular and does not result in a single reflection, but probably, for a very large portion of the sound, of multiple reflection before it finally emerges. The physical conditions are such that they obviously do not admit of analytic expression, but the explanation of the great absorption by an extended audience surface is not difficult to understand. In addition to the above there is another partial explanation which contributes to the results. The felt forms a perfectly continuous medium, and therefore offers a comparatively rigid reflecting surface. The comparatively light, thin, and porous nature of the clothing of women, perhaps more than of men, contributes to the great absorption of the high notes.

The next experiment, taking them up chronologically, and perhaps next even from the standpoint of interest, was in regard to a brick wall-surface. This experiment was carried out in the constant-temperature room mentioned in the previous papers. The arrangement of apparatus is shown in Fig. 3, where the air reservoir in the room above is shown in dotted lines. In many respects the constant-temperature room offered admirable conditions for the experiment.

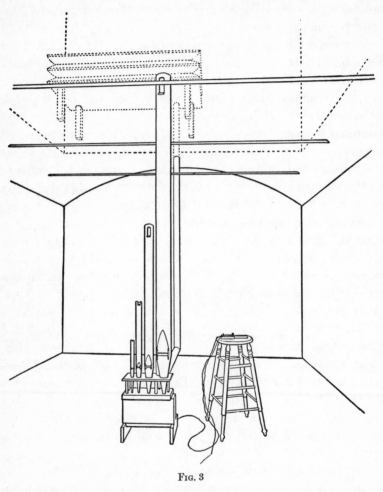

Fig. 3

Its position in the center of the building and its depth underground made it comparatively free from outside disturbing noises, — so much so that it was possible to experiment in this room in the earlier parts of the evening, although not, of course, when any one else was at work in the building. While it possesses these advantages, its arched ceiling, by placing it in the category of special cases, makes extra precaution necessary. Fortunately, at the beginning of the experiment the walls were unpainted. Under these conditions its

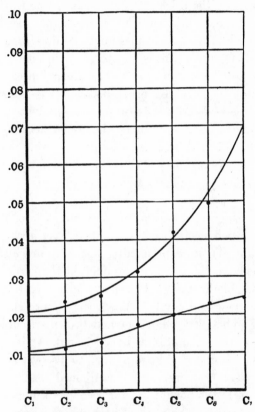

FIG. 4. The absorbing power of a 45 cm. thick brick wall. The upper curve represents the absorbing power of an unpainted brick surface. The bricks were hard but not glazed, and were set in cement. The lower curve represents the absorbing power of the same surface painted with two coats of oil paint. The difference between the two curves represents the absorption due to the porosity of the bricks. In small part, but probably only in small part, the difference is due to difference in superficial smoothness. C_3 (middle C) 256.

coefficient of absorption for different notes was determined. It was then painted with an oil paint, two coats, and its coefficient of absorption redetermined. The two curves are shown in Fig. 4. The upper curve is for the unpainted brick; the lower curve is that obtained after the walls were painted. The difference between the two curves would, if plotted alone, be the curve of absorption due to the porosity of the brick. It may seem, perhaps, that the paint in covering the bare brick wall made a smoother surface, and the difference between the two results might be due in part to less surface friction. Of course this is a factor, but that it is an exceedingly small factor will be shown later in the discussion of the results on the absorption of sound by other bodies. The absorption of the sound after the walls are painted is, of course, due to the yielding of the walls under the vibration, to the sound actually transmitted bodily by the walls, and to the absorption in the process of transmission. It is necessary to call attention to the fact that the vertical ordinates are here magnified tenfold over the ordinates shown in the last curve.

The next experiment was on the determination of the absorption of sound by wood sheathing. It is not an easy matter to find conditions suitable for this experiment. The room in which the absorption by wood sheathing was determined in the earlier experiments was not available for these. It was available then only because the building was new and empty. When these more elaborate experiments were under way the room had become occupied, and in a manner that did not admit of its being cleared. Quite a little searching in the neighborhood of Boston failed to discover an entirely suitable room. The best one available adjoined a night lunch room. The night lunch was bought out for a couple of nights, and the experiment was tried. The work of both nights was much disturbed. The traffic past the building did not stop until nearly two o'clock, and began again about four. The interest of those passing by on foot throughout the night, and the necessity of repeated explanations to the police, greatly interfered with the work. This detailed statement of the conditions under which the experiment was tried is made by way of explanation of the irregularity of the observations recorded on the curve, and of the failure to carry this particular line of work further. The first night seven points were obtained for the seven notes C_1 64 to C_7 4096. This work was done by means of a portable apparatus shown in Fig. 5. The reduction of these results on the following day showed variations indicative of maxima and minima, which to be accurately located would require the de-

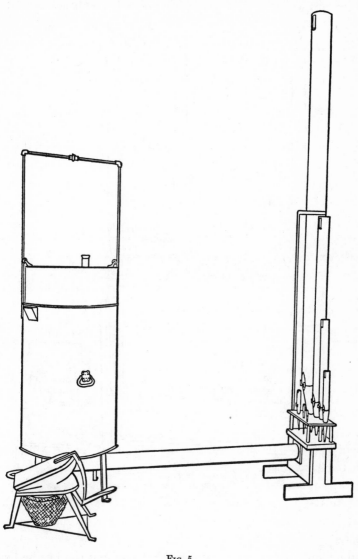

Fig. 5

termination of intermediate points. The experiment the following night was by means of the organ shown in Fig. 6, and points were determined for the E and G notes in each octave between C_2 128 and C_6 2048. Other points would have been determined, but time did not permit. It is obvious that the intermediate points in the lower and

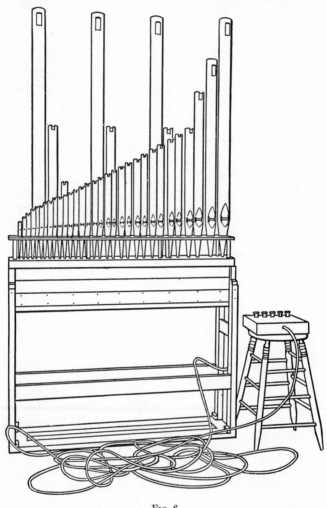

Fig. 6

in the higher octave were desirable, but no pipes were to be had on such short notice for this part of the range, and in their absence the data could not be obtained. In the diagram, Fig. 7, the points lying on the vertical lines were determined the first night. The points lying between the vertical lines were determined the second night. The accuracy with which these points fall on a smooth curve is

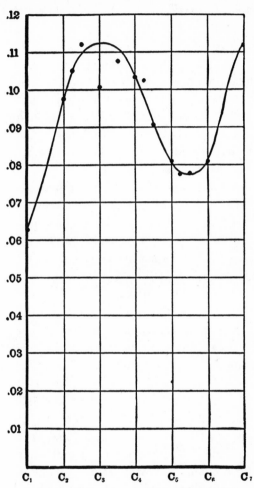

FIG. 7. The absorbing power of wood sheathing, two centimeters thick, North Carolina pine. The observations
were made under very unsuitable conditions. The
absorption is here due almost wholly to yielding of the
sheathing as a whole, the surface being shellacked,
smooth, and non-porous. The curve shows one point
of resonance within the range tested, and the probability of another point of resonance above. It is not
possible now to learn as much in regard to the framing
and arrangement of the studding in the particular room
tested as is desirable. C_3 (middle C) 256.

perhaps all that could be expected in view of the difficulty under which the observations were conducted and the limited time available. One point in particular falls far off from this curve, the point for C_3 256, by an amount which is, to say the least, serious, and which can be justified only by the conditions under which the work was done. The general trend of the curve seems, however, established beyond reasonable doubt. It is interesting to note that there is one point of maximum absorption, which is due to resonance between the walls and the sound, and that this point of maximum absorption lies in the lower part, though not in the lowest part, of the range of pitch tested. It would have been interesting to determine, had the time and facilities permitted, the shape of the curve beyond C_7 4096, and to see if it rises indefinitely, or shows, as is far more likely, a succession of maxima. The scale employed in this curve is the same as that employed in the diagram of the unpainted and painted wall-surfaces. It may perhaps be noted in this connection that at the very least the absorption is four times that of painted brick walls.

The experiment was then directed to the determination of the absorption of sound by cushions, and for this purpose return was made to the constant-temperature room. Working in the manner indicated in the earlier papers for substances which could be carried in and out of a room, the curves represented in Fig. 8 were obtained. Curve 1 shows the absorption coefficient for the Sanders Theatre cushions, with which the whole investigation was begun ten years ago. These cushions were of a particularly open grade of packing, a sort of wiry grass or vegetable fiber. They were covered with canvas ticking, and that in turn with a very thin cloth covering. Curve 2 is for cushions borrowed from the Phillips Brooks House. They were of a high grade, filled with long curly hair, and covered with canvas ticking, which was in turn covered by a long nap plush. Curve 3 is for the cushions of Appleton Chapel, hair covered with a leatherette, and showing a sharper maximum and a more rapid diminution in absorption for the higher frequencies, as would be expected under such conditions. Curve 4 is probably the most interesting, because for more standard commercial conditions. It is the curve for elastic felt cushions as made by Sperry and Beale.

It is to be observed that all four curves fall off for the higher frequencies, all show a maximum located within an octave, and three

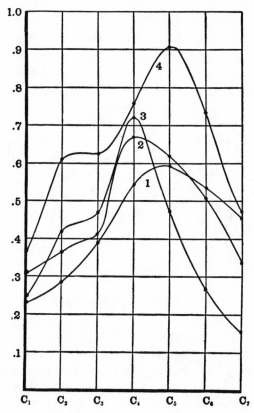

Fig. 8. The absorbing power of cushions. Curve 1 is for "Sanders Theatre" cushions of wiry vegetable fiber, covered with canvas ticking and a thin cloth. Curve 2 is for "Brooks House" cushions of long hair, covered with the same kind of ticking and plush. Curve 3 is for "Appleton Chapel" cushions of hair, covered with ticking and a thin leatherette. Curve 4 is for the elastic felt cushions of commerce, of elastic cotton, covered with ticking and short nap plush. The absorbing power is per square meter of surface. C_3 (middle C) 256.

of the curves show a curious hump in the second octave. This break in the curve is a genuine phenomenon, as it was tested time after time. It is perhaps due to a secondary resonance, and it is to be observed that it is the more pronounced in those curves that have the sharper resonance in their principal maxima.

Observations were then obtained on unupholstered chairs and settees. The result for chairs is shown in Fig. 10. This curve gives the absorption coefficient per single chair. The effect was surprisingly small; in fact, when the floor of the constant-temperature room was entirely covered with the chairs spaced at usual seating distances, the effect on the reverberation in the room was exceed-

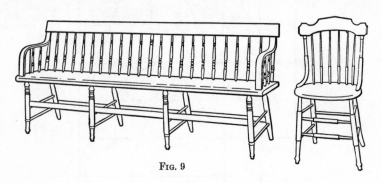

Fig. 9

ingly slight. The fact that it was so slight and the consequent difficulty in measuring the coefficient is a partial explanation of the variation of the results as indicated in the figure. Nevertheless it is probable that the variations there indicated have some real basis, for a repetition of the work showed the points again falling above and below the line as in the first experiment. The amount that these fell above and below the line was difficult to determine, and the number of points along the curve were too few to justify attempting to follow their values by the line. In fact the line is drawn on the diagram merely to indicate in a general way the fact that the coefficient of absorption is nearly the same over the whole range. A varying resonance phenomenon was unquestionably present, but so small as to be negligible; and in fact the whole absorption by the chairs is an exceedingly small factor. The chair was of ash, and its type is shown in the accompanying sketch, Fig. 9.

The results of the observations on settees is shown in Fig. 11. Those plotted are the coefficients per single seat, there being five seats to the settee. The settees were placed at the customary distance. Here again the principal interest attaches to the fact that the coefficient of absorption is so exceedingly small that the total effect on the reverberation is hardly noticeable. Here also the plotted results do not fall on the line drawn, and the departure is

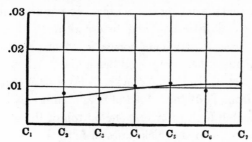

FIG. 10. The absorbing power of ash chairs shown in Fig. 9.

due probably to some slight resonance. The magnitude of the departure, however, could not be determined with accuracy because of the small magnitude of the total absorption coefficient. For these reasons and because the number of points was insufficient, no at-

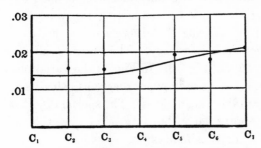

FIG. 11. The absorbing power of ash settees shown in Fig. 9. The absorption is per single seat, the settee as shown seating five.

tempt was made to draw the curve through the plotted points, but merely to indicate a plotted tendency. The settees were of ash, and their general style is shown in the sketch.

An investigation was then begun in regard to the nature of the process of absorption of sound. The material chosen for this work was a very durable grade of felt, which, as the manufacturers claimed, was all wool. Even a casual examination of its texture makes it difficult to believe that it is all wool. It has, however, the advantage of being porous, flexible, and very durable. Almost constant handling for several years has apparently not greatly changed its consistency. It is to be noted that this felt is not that mentioned in the papers of six years ago. That felt was of lime-treated cow's hair, the kind used in packing steam pipes. It was very much cheaper in price, but stood little handling before disintegrating. The felt employed in these experiments comes in sheets of various thicknesses, the thickness here employed being about 1.1 cm.

The coefficient of absorption of a single layer of felt was measured for the notes from C_1 64 to C_7 4096 at octave intervals. The experiment was repeated for two layers, one on top of the other, then for three, and so on up to six thicknesses of felt. Because the greater thicknesses presented an area on the edge not inconsiderable in comparison with the surface, the felt was surrounded by a narrow wood frame. Under such circumstances it was safe to assume that the absorption was entirely by the upper surface of the felt. The experiment was repeated a great many times, first measuring the coefficient of absorption for one thickness for all frequencies, and then checking the work by conducting experiments in the other order; that is, measuring the absorption by one, two, three, etc., thicknesses, for each frequency. The mean of all observations is shown in Fig. 12 and Fig. 13. In Fig. 12 the variations in pitch are plotted as abscissas, as in previous diagrams, whereas in Fig. 13 the thicknesses are taken as abscissas. The special object of the second method will appear later, but a general object of adopting this method of plotting is as follows:

If we consider Fig. 12, for example, the drawing of the line through any one set of points should be made not merely to best fit those points, but should be drawn having in mind the fact that it, as a curve, is one of a family of curves, and that it should be drawn not merely as a best curve through its own points, but as best fits the whole set. For example, in Fig. 12 the curve for four thicknesses would not have been drawn as there shown if drawn simply with reference to its own points. It would have been drawn directly through the points for C_1 64 and C_2 128. Similarly the curve for five thicknesses would have been drawn a little nearer the point for C_2 128, and above instead of below the point for C_1 64. Considering, however, the whole family of curves and recognizing that each point is not without some error, the curves as drawn are more nearly correct. The best method of reconciling the several curves to each

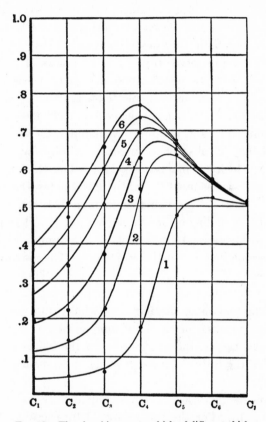

Fig. 12. The absorbing power of felt of different thicknesses. Each piece of felt was 1.1 cm. in thickness. Curve 1 is for a single thickness, curve 2 for two thicknesses placed one on top of the other, etc. As shown by these curves, the absorption is in part by penetration into the pores of the felt, in part by a yielding of the mass as a whole. Resonance in the latter process is clearly shown by a maximum shifting to lower and lower pitch with increase in thickness of the felt. C_3 (middle C) 256.

other is to plot two diagrams, one in which the variations in pitch are taken as abscissa and one in which the variations in thickness of felt are taken as abscissas; then draw through the points the best fitting curves and average the corresponding ordinates taken from the curves thus drawn; and with these average ordinates redraw both families of curves. The points shown on the diagram are of course the original results obtained experimentally. In general they fall pretty close to the curves, although at times, as in the points noted, they fall rather far to one side.

The following will serve to present the points of particular interest revealed by the family of curves in Fig. 12, where the absorption by the several thicknesses is plotted against pitch for abscissas. It is to be observed that a single thickness scarcely absorbs the sound from the eight, four, and two-foot organ pipes, C_1 64, C_2 128, and C_3 256, and that its absorption increases rapidly for the next two octaves, after which it remains a constant. Two thicknesses absorb more — about twice as much — for the lower notes, the curve rising more rapidly, passing through a maximum between C_4 512 and C_5 1024, and then falling off for the higher notes. The same is true for greater thicknesses. All curves show a maximum, each succeeding one corresponding to a little lower note. The maximum for six thicknesses coincides pretty closely to C_4 512. The absorption of the sound by felt may be ascribed to three causes, — porosity of structure, compression of the felt as a whole, and friction on the surface. The presence of the maximum must be ascribed to the second of these causes, the compression of the felt as a whole. As to the third of these three causes, it is best to consult the curves of the next figure.

The following facts are rendered particularly evident by the curves of Fig. 13. For the tones emitted by the eight-foot organ pipe, C_1 64, the absorption of the sound is very nearly proportional to the thickness of the felt over the range tested, six thicknesses, 6.6 cm. The curves for notes of increasing pitch show increasing value for the coefficients of absorption. They all show that were the thickness of the felt sufficiently great, a limit would be approached — a fact, of course, self-evident — but for C_5 1024 this thickness was reached within the range experimented on; and of course the same is true for all higher notes, C_6 2048 and C_7 4096. The higher the note, the less the thickness of felt necessary to produce a maximum effect. The curves of C_1 64, C_2 128, C_3 256, and

C_4 512, if extended backward, would pass nearly through the origin. This indicates that for at least notes of so low a pitch the absorption

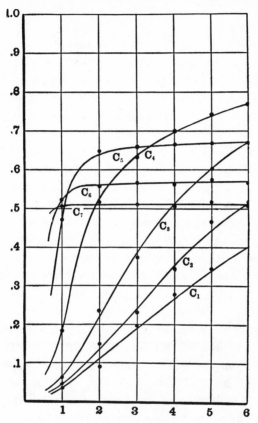

Fig. 13. The absorbing power of felt of different thicknesses. The data, Fig. 12, is here plotted in a slightly different manner — horizontally on plotted increasing thickness — and the curves are for notes of different frequency at octave intervals in pitch. Thus plotted the curves show the necessary thickness of felt for practically maximum efficiency in absorbing sound of different pitch. These curves also show that for the lowest three notes surface friction is negligible, at least in comparison with the other factors. For the high notes one thickness of felt was too great for the curves to be conclusive in regard to this point. C_3 (middle C) 256.

of sound would be zero, or nearly zero, for zero thickness. Since zero thickness would leave surface effects, the argument leads to the conclusion that surface friction as an agent in the absorption of sound is of small importance. The curves plotted do not give any evidence in this respect in regard to the higher notes, C_5 1024, C_6 2048, and C_7 4096.

It is of course evident that the above data do not by any means cover all the ground that should be covered. It is highly desirable that data should be accessible for glass surfaces, for glazed tile surfaces, for plastered and unplastered porous tile, for plaster on wood lath and plaster on wire lath, for rugs and carpets; but even with these data collected the job would be by no means completed. What is wanted is not merely the measurement of existing material and wall-surfaces, but an investigation of all the possibilities. A concrete case will perhaps illustrate this. If the wall-surface is to be of wood, there enter the questions as to what would be the effect of varying the material, — how ash differs from oak, and oak from walnut or pine or whitewood; what is the effect of variations in thickness; what the effect of paneling; what is the effect of the spacing of the furring on which the wood sheathing is fastened. If the wall is to be plaster on lath, there arises the question as to the difference between wood lath and wire lath, between the mortar that was formerly used and the wall of today, which is made of hard and impervious plaster. What is the effect of variations in thickness of the plaster? What is the effect of painting the plaster in oil or in water colors? What is the effect of the depth of the air space behind the plaster? The recent efforts at fireproof construction have resulted in the use of harder and harder wall-surfaces, and great reverberation in the room, and in many cases in poorer acoustics. Is it possible to devise a material which shall satisfy the conditions as to fireproof qualities and yet retain the excellence of some of the older but not fireproof rooms? Or, if one turns to the interior furnishings, what type of chair is best, what form of cushions, or what form of upholstery? There are many forms of auditorium chairs and settees, and all these should be investigated if one proposes to apply exact calculation to the problem. These are some of the questions that have arisen. A few data have been obtained looking toward the answer to some of them. The difficulty in the way of the prosecution of such work is greater, however, than appears at first sight, the particular difficulties being of opportunity and of expense. It is difficult, for example, to find rooms whose walls are in large measure of glass, especially when one bears in mind that the room must be empty, that its other wall-surfaces must be of a substance fully investigated, and that it must be in a location admitting of quiet work. Or, to investigate the effect of the different kinds of plaster and of the different methods of plastering, it is necessary to have a room, preferably an underground room, which can be lined and relined. The constant-temperature room which is now available for the experiments is not a room suitable to that particular investigation, and for best results a special room should be constructed. Moreover, the expense of plastering and replastering a room — and this process, to arrive at anything like a general solution of the problem, would have to be done a great many times — would be very great, and is at the present moment prohibitive. A little data along some of these lines have been secured, but not at all in final form. The work in the past has been largely of an analytical nature. Could the investigation take the form of constructive research, and lead to new methods and greater possibilities, it would be taking its more interesting form.

The above discussion has been solely with reference to the determination of the coefficient of absorption of sound. It is now proposed to discuss the question of the application of these coefficients to the calculation of reverberation. In the first series of papers, reverberation was defined with reference to C_4 512 as the continuation of the sound in a room after the source had ceased, the initial intensity of the sound being one million times minimum audible intensity. It is debatable whether or not this definition should be extended without alteration to reverberation for other notes than C_4 512. There is a good deal to be said both for and against its retention. The whole, however, hinges on the outcome of a physiological or psychological inquiry not yet in such shape as to lead to a final decision. The question is therefore held in abeyance, and for the time the definition is retained.

Retaining the definition, the reverberation for any pitch can be calculated by the formula

$$T = \frac{KV}{a},$$

where V is the volume of the room, K is a constant depending on the initial intensity, and a is the total absorbing power of the walls and the contained material. K and V are the same for all pitch

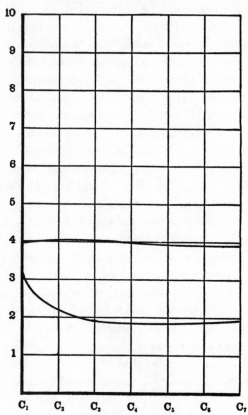

Fig. 14. Curves expressing the reverberation in the large lecture-room of the Jefferson Physical Laboratory with (lower curve) and without (upper curve) an audience. These curves express in seconds the duration of the residual sound in the room after the cessation of sources producing intensities 10^6 times minimum audible intensity for each note. The upper curve describes acoustical conditions which are very unsatisfactory, as the hall is to be used for speaking purposes. The lower curve describes acoustically satisfactory conditions. C_3 (middle C) 256.

frequencies. K is .164 for an initial intensity 10^6 times minimum audible intensity. The only factor that varies with the pitch is a, which can be determined from the data given above.

In illustration, the curves in the accompanying Fig. 14 give the reverberation in the large lecture-room of the Jefferson Physical Laboratory. The upper curve defines the reverberation in the room when entirely empty; the lower curve defines this reverberation in the same room with an audience two-thirds filling the room. The upper curve represents a condition which would be entirely impractical for speaking purposes; the lower curve represents a fairly satisfactory condition.

3

MELODY AND THE ORIGIN OF THE MUSICAL SCALE [1]

In the vice-presidential addresses of the American Association great latitude in the choice of subjects is allowed and taken, but there is, I believe, no precedent for choosing the review of a book printed fifty-five years before. Helmholtz' *Tonenemfindungen*, produced by a masterful knowledge of physiology, physics, and mathematics, and a scholar's knowledge of the literature of music, has warded off all essential criticism by its breadth, completeness, and wealth of detail. Since it was first published it has been added to by the author from time to time in successive editions, and greatly bulwarked by the scholarly notes and appendices of its translator, Dr. Alexander J. Ellis. The original text remains unchanged, and unchallenged, as far as physicists are concerned, in all important respects. In taking exception at this late day to the fundamental thesis of Part III, I derive the necessary courage from the fact that should such exception be sustained, it will serve to restore to its full application that greater and more original contribution of Helmholtz which he included in Part II. Having given a physical and physiological explanation of the harmony and discord of simultaneous sounds, and, therefore, an explanation of the musical scale as used in modern composition, Helmholtz was met by an apparent anachronism. The musical scale, identical with the modern musical scale in all essentials, antedated by its use in single-part melody the invention of chordal composition, or, as Helmholtz expressed it, preceded all experience of musical harmony. In seeking an explanation of this early invention of the musical scale, Helmholtz abandoned his most notable contribution, and relegated his explanation of harmony and discord to the minor service of explaining a fortunate, though of course an important use of an already invented system of musical notes. The explanation of the original

[1] Vice-Presidential Address, Section B, American Association for the Advancement of Science, Chicago, 1907.

invention of the musical scale and its use in single-part music through the classical and the early Christian eras, he sought for in purely aesthetic considerations, — in exactly those devices from which he had just succeeded in rescuing the explanation of harmony and discord.

The human ear consists of three parts, — in the nomenclature of anatomy, of the outer, middle, and inner ear. The outer and the inner ears are connected by a series of three small bones traversing the middle ear and transmitting the vibrations of sound. The inner ear is a peculiarly shaped cavity in one of the hard bones of the skull. That part of the cavity with which we are here concerned is a long passage called from its resemblance to the interior of a snail shell the cochlea. The cavity has two windows which are closed by membranes. It is to the uppermost of these membranes that the train of three small bones, reaching from the drum of the outer ear, is attached at its inner end. It is to this upper membrane, therefore, that the vibration is communicated, and through it the vibration reaches the fluid which fills the inner cavity. As the membrane covering the upper window vibrates, the membrane covering the lower window yielding, also vibrates, and the motion of the fluid is in the nature of a slight displacement from one to the other window, to and fro. From between these windows a diaphragm, dividing the passageway, extends almost the whole length of the cochlea. This diaphragm is composed in part of a great number of very fine fibers stretched side by side, transverse to the cochlea, and called after their discoverer, fibers of Corti. On this diaphragm terminate the auditory nerves. When the liquid vibrates, the fibers vibrate in unison, the nerve terminals are stimulated, and thus the sensation of sound is produced. These fibers of Corti are of different lengths and presumably are stretched with different tensions. They therefore have different natural rates of vibration and a sympathetic resonance for different notes. The whole has been called a harp of several thousand strings.

Were these fibers of Corti very free in their vibration, each would respond to and would respond strongly only to that particular note with whose frequency it is in unison. Because of the fact that they are in a liquid, and possibly also because of the manner of their terminal connections, they are considerably damped. Because of this their response is both less in amount and less selective in character. In fact, under these conditions, not one, but many fibers vibrate in response to a single pure note. A considerable length or area of the diaphragm is excited. So long as the exciting sound remains pure in quality, constant in pitch, and constant in intensity, the area of the diaphragm affected and the amplitude of its vibration remain unchanged. If, however, two notes are sounded of nearly the same pitch, the areas of the diaphragm affected by the two notes overlap. In the overlapping region the vibration is violent when the two notes are in the same phase, weak when they are in opposite phase. The result is the familiar phenomena of beats. Such beats when slow are not disagreeable and not without musical value. If the difference between the two notes is increased, the beats become more rapid and more disagreeable. To this violent disturbance, to the starting and stopping of the vibration of the fibers of Corti, Helmholtz ascribed the sense of roughness which we call discord. As the notes are more widely separated in pitch, the overlapping of the affected areas diminishes. Between pure notes the sense of discord disappears with sufficient separation in pitch. When the two vibrating areas exactly match, because the two notes are of exactly the same pitch, and when the two areas do not in the least overlap, because of a sufficiently wide separation in pitch, the result according to Helmholtz is harmony. Partial overlapping of the affected areas produces beats, and the roughness of beats is discord. Such, reduced to its fewest elements, is Helmholtz' explanation of the harmony and discord of tones which are pure.

But no musical tone is simple. It always consists of a combination of so-called partial tones which bear to each other a more or less simple relationship. Of these partial tones, one is called the fundamental, — so-called because it is the loudest or lowest or, better still, because it is that to which the other partial tones bear the simplest relation. A musical tone, therefore, affects not one, but, through its fundamental and upper partial tones, several areas of the diaphragm in the cochlea. Two musical tones, each with its fundamental and upper partials, therefore, affect areas of the diaphragm which overlap each other in a more or less complicated manner, depending on the relative frequencies of the fundamental tones and the relationships of their upper partials. The exact matching of the areas affected by these two systems of partial tones, or the entire separation of the affected areas, give harmony. The overlapping of these affected areas, if great, produces discord, or, if slight in amount, modifications and color of harmony.

In the great majority of musical tones the upper partials bear simple relationships to the fundamentals, being integral multiples in vibration frequency. Helmholtz showed that if of two such tones one continued to sound unchanged in pitch, and the other starting in unison was gradually raised in pitch, the resulting discord would pass through maxima and minima, and that the minima would locate the notes of the pentatonic scale. The intermediate notes of the complete modern musical scale are determined by a repetition of this process starting from the notes thus determined.

If to this is added a similar consideration of the mutual interference of the combinational tones which are themselves due to the interaction of the partial tones, we have the whole, though of course in the briefest outline, of Helmholtz' theory of the harmony and discord of simultaneously sounding musical tones.

Having thus in Parts I and II developed a theory for the harmony and discord of simultaneous sounds, and having developed a theory which explains the modern use of the musical scale in chords and harmonic music, Helmholtz pointed out, in Part III, that the musical scale in its present form existed before the invention of harmonic music and before the use of chords.

Music may be divided into three principal periods: —

1. "Homophonic or Unison Music of the ancients," including the music of the Christian era up to the eleventh century, "to which also belongs the existing music of Oriental and Asiatic nations."

2. "Polyphonic music of the middle ages, with several parts, but without regard to any independent musical significance of the harmonies, extending from the tenth to the seventeenth century."

3. "Harmonic or modern music characterized by the independent significance attributed to the harmonies as such."

Polyphonic music was the first to call for the production of simultaneous sounds, and, therefore, for the hearing or the experience of musical harmony. Homophonic music, that which alone existed up to the tenth or eleventh century, consisted in the progression of single-part melody. Struck by this fact, Helmholtz recognized the necessity of seeking another explanation for the invention and the use of a scale of fixed notes in the music of this period. To borrow his own words, "scales existed long before there was any knowledge or experience of harmony." Again, elsewhere, he says in emphasizing the point: "The individual parts of melody reach the ear in succession. We cannot perceive them all at once; we cannot observe backwards and forwards at pleasure." Between sounds produced and heard in discrete succession, there can be neither harmony nor discord, there cannot be beats, or roughness or interruption of continuous vibrations. Regarding the sounds of a melody as not merely written in strict and non-overlapping succession, but also as produced and heard in discrete succession, Helmholtz sought another basis for the choice of the notes to constitute a scale for homophonic music. His explanation of this invention can be best presented by a few quotations: —

Melody has to express a motion in such a manner that the hearer may easily, clearly, and certainly appreciate the character of that motion by immediate perception. This is only possible when the steps of this motion, their rapidity, and their amount, are also exactly measurable by immediate sensible perception. Melodic motion is change of pitch in time. To measure it perfectly, the length of time elapsed and the distance between the pitches must be measurable. This is possible for immediate audition only on condition that the alterations both in time and pitch should proceed by regular and determinate degrees.

Again Helmholtz says: —

For a clear and sure measurement of the change of pitch no means was left but progression by determinate degrees. This series of degrees is laid down in the musical scale. When the wind howls and its pitch rises or falls in insensible gradations without any break, we have nothing to measure the variations of pitch, nothing by which we can compare the later with the earlier sounds, and comprehend the extent of the change. The whole phenomenon produces a confused, unpleasant impression. The musical scale is as it were the divided rod, by which we measure progression in pitch, as rhythm measures progression in time.

Later he says: —

Let us begin with the Octave, in which the relationship to the fundamental tone is most remarkable. Let any melody be executed on any instrument which has a good musical quality of tone, such as a human voice; the hearer must have heard not only the primes of the compound tones, but also their upper octaves, and, less strongly, the remaining upper partials. When, then, a higher voice afterwards executes the same melody an Octave higher, we hear again a part of what we heard before, namely the evenly numbered partial tones of the former compound tones, and at the same time we hear nothing that we had not previously heard.

What is true of the Octave is true in a less degree for the Twelfth. If a melody is repeated in the Twelfth we again hear only what we had already heard, but the repeated part of what we heard is much weaker, because only the third, sixth, ninth, etc., partial tone is repeated, whereas for repetition in the Octave, instead of the third partial, the much stronger second and weaker fourth partial is heard, and in place of the ninth, the eighth and tenth occur, etc.

For the repetition on the Fifth, only a part of the new sound is identical with a part of what had been heard, but it is, nevertheless, the most perfect repetition which can be executed at a smaller interval than an Octave.

Without carrying these quotations further they will suffice to illustrate the basis which Helmholtz would ascribe to homophonic music and early melodic composition. On this explanation the basis of melody is purely that of rhythm and rhythm based on a scale of intervals. The scale of intervals in turn is based on a recognition, conscious or subconscious, of the compound character of musical tones, and of the existence in tones of different pitch of partials of the same pitch. This calls for a degree of musical insight and discrimination which it is difficult to credit to a primitive art. It is in reality the skill of the highly trained musician, of a musician trained by long experience with sounds which are rich and accurate in quality. This power of analysis goes rather with supreme skill than with the early gropings of an art.

After having developed a theory of harmony and discord based on elaborate experimental and mathematical investigations, which was remarkable in bringing together three such diverse fields as physics, physiology, and aesthetics, he relegated it to the minor application of explaining the use in modern music of an already existing and highly developed musical scale, and sought an explanation of the earlier use of the scale in melody and its original invention in the principle which is very far from possessing either the beauty or the convincing quality of his earlier hypothesis. He was forced to this by a priority of melodic or homophonic composition. He saw in melody only a succession of notes, no two existing at the same time, and therefore incapable of producing harmony or discord in a manner such as he had been considering.

It is true that melody is written as a pure succession of discrete notes, one beginning only when the other has ceased. It is true also that melody is so sung and so produced on a homophonic instrument, such as the voice, flute, reeds, or one-stringed instruments. This is peculiarly true of the voice, and it is with the voice that one naturally associates the earliest invention of the scale. But while it is true that the earliest song must have consisted of tones produced only in succession, it is not necessarily true that such sounds were heard as isolated notes. A sound produced in a space which is in any way confined continues until it is diminished by transmission through openings or is absorbed by the retaining walls, or contained material to such a point that it is past the threshold of audibility, and this prolongation of audibility of sound is under many conditions a factor of no inconsiderable importance. In many rooms of ordinary construction the prolongation of audibility amounts to two or three seconds, and it is not exceedingly rare that a sound of moderate initial intensity should continue audible for eight, nine, or even ten seconds after the source has ceased. As a result of this, single-part music produced as successive separate sounds is, nevertheless, heard as overlapping, and at times as greatly overlapping tones. Each note may well be audible with appreciable intensity not merely through the next, but through several succeeding notes. Under such conditions we have every opportunity, even with single-part music, for the production of all the phenomena of harmony and discord which has been discussed by Helmholtz in explanation of the chordal use of the musical scale. In any ordinarily bare and uncarpeted room, one may sing in succession a series of notes and then hear for some time afterward their full chordal effect.

All the arguments that Helmholtz advanced in support of his hypothesis, that the musical scale was devised solely from considerations of rhythm and founded on a repetition of faint upper partials, hold with equal force in the explanation here proposed. The identity of partial tones in compound tones with different fundamentals is one of the conditions of harmony, and the scale devised by considerations of the mutual harmony of the notes sounded simultaneously, would, in every respect, be the same as that of a scale based on repeated upper partials. In the one case the identity of upper partials is an act of memory, in the other it is determined by the harmony of sustained tones. All the arguments by Helmholtz based on historical considerations and on racial and national differences are equally applicable to the hypothesis of sustained tones. In fact, they take on an additional significance, for we may now view all these differences not merely in the light of differences in racial development and temperament, but in the light of physical environment. Housed or unhoused, dwelling in reed huts or in tents, in houses of wood or of stone, in houses and temples high vaulted or low roofed, of heavy furnishing or light, in these conditions we may look for the factors which determine the development of a musical scale in any race, which determine the rapidity of the growth of the scale, its richness, and its considerable use in single-part melody.

The duration of audibility of a sound depends on its initial intensity and on its pitch, to a small degree on the shape of the confined space, and to a very large degree on the volume of the space and on the material of which the walls are composed. The duration of audibility is a logarithmic function of the initial intensity, and as the latter is practically always a large multiple of the minimum audible intensity, this feature of the problem may be neglected when considering it broadly. For this discussion we may also leave out of consideration the effect of shape as being both minor and too intricately variable. The pitch here considered will be the middle of the musical scale; for the extremes of the scale the figures would be very different. The problem then may be reduced to two factors, volume and material. It is easy to dispose of the problem reduced to these two elements.

The duration of audibility of a sound is directly proportional to the volume of a room and inversely proportional to the total absorbing power of the walls and the contained material. The volume of the room, the shape remaining the same, is proportional to the cube, while the area of the walls is proportional to the square of the linear dimensions. The duration of audibility, proportional to the ratio of these two, is proportional to the first power of the linear dimension. Other things being equal, the duration of audibility, the overlapping of successive sounds, and, therefore, the experience of harmony in single-part music is proportional to the linear dimensions of the room, be it dwelling house or temple.

Turning to the question of material the following figures are suggestive: Any opening into the outside space, provided that outside space is itself unconfined, may be regarded as being totally absorbing. The absorbing power of hard pine wood sheathing of one-half inch thickness is 6.1 per cent; of plaster on wood lath, 3.4 per cent; of single-thickness glass, 2.7 per cent; of brick in Portland cement, 2.5 per cent; of the same brick painted with oil paint, 1.4 per cent. Wood sheathing is nearly double any of the rest. On the other hand, a man in the ordinary clothing of today is equal in his absorbing power to nearly 48 per cent of that of a square meter of unobstructed opening, a woman is 54 per cent, and a square meter of audience at ordinary seating distance is nearly 96 per cent. Of significance also in this connection is the fact that Oriental rugs have an absorbing power of nearly 29 per cent, and house plants of 11 per cent.

Of course, the direct application of these figures in any accurate calculation of the conditions of life among different races or at different periods of time is impossible, but they indicate in no uncertain manner the great differences acoustically in the environment of Asiatic races, of aboriginal races in central and southern Africa, of the Mediterranean countries, of northern Europe at different periods of time. We have explained for us by these figures why the musical scale has but slowly developed in the greater part of Asia and of Africa. Almost no traveler has reported a musical scale, even of the most primitive sort, among any of the previously unvisited tribes of Africa. This fact could not be ascribed to racial inaptitude. If melody was, as Helmholtz suggested, but rhythm in time and in pitch, the musical scale should have been developed in Africa if anywhere. These races were given to the most rhythmical dancing, and the rhythmical beating of drums and tomtoms. Rhythm in time they certainly had. Moreover, failure to develop a musical scale could not be ascribed to racial inaptitude to feeling for pitch. Transported to America and brought in contact with the musical scale, the negro became immediately the most musical part of our population. The absence of a highly developed scale in Africa must then be ascribed to environment.

Turning to Europe we find the musical scale most rapidly developing among the stone-dwelling people along the shores of the Mediterranean. The development of the scale and its increased use kept pace with the increased size of the dwellings and temples. It showed above all in their religious worship, as their temples and churches reached cathedral size. The reverberation which accompanied the lofty and magnificent architecture increased until even the spoken service became intoned in the Gregorian chant. It is not going beyond the bounds of reason to say that in those churches in Europe which are housed in magnificent cathedrals, the Catholic, the Lutheran, and Protestant Episcopal, the form of worship is in part determined by their acoustical conditions.

This presents a tempting opportunity to enlarge on the fact that the alleged earliest evidence of a musical scale, a supposed flute, belonged to the cave dwellers of Europe. This and the impulse to sing in an empty room, and the ease with which even the unmusical can keep the key in simple airs under such conditions, are significant facts, but gain nothing by amplification. The same may be said of the fact that since music has been written for more crowded auditoriums and with harmonic accompaniment melody has become of less harmonious sequence. These and many other instances of the effect of reverberation come to mind.

In conclusion, it may not be out of place to repeat the thesis that melody may be regarded not only as rhythm in time and rhythm in pitch, but also as harmony in sustained tones, and that we may see in the history of music, certainly in its early beginnings, but possibly also in its subsequent development, not only genius and invention, but also the effect of physical environment.

4

ARCHITECTURAL ACOUSTICS[1]

EFFECTS OF AIR CURRENTS AND OF TEMPERATURE

ORDINARILY there is not a close connection between the flow of air in a room and its acoustical properties, although it has been frequently suggested that thus the sound may be carried effectively to different parts. On the other hand, while the motion of the air is of minor importance, the distribution of temperature is of more importance, and it is on reliable record that serious acoustical difficulty has arisen from abrupt differences of temperature in an auditorium. Finally, transmission of disturbing noises through the ventilation ducts, perhaps theoretically a side issue, is practically a legitimate and necessary part of the subject. The discussion will be under these three heads.

The first of the above three topics, the possible effect of the motion of the air on the acoustical property of a room, is the immediate subject.

VENTILATION

It was suggested during the planning of the Boston Symphony Hall that its acoustical properties would be greatly benefited by introducing the air for ventilation at the front and exhausting at the back, thus carrying the sound by the motion of the air the length of the room. The same suggestion has been made to the writer by others in regard to other buildings, but this case will serve as sufficient example. The suggestion was unofficial and the gentleman proposing it accompanied it by a section of a very different hall from the hall designed by Mr. McKim, but as this section was only a sketch and without dimensions the following calculation will be made as if the idea were to be applied to the present hall. It will be shown that the result thus to be secured, while in the right

---

[1] Engineering Record, June, 1910.

direction, is of a magnitude too small to be appreciable. To make this the more decisive we shall assume throughout the argument the most favorable conditions possible.

If a sound is produced in still air in open space it spreads in a spherical wave diminishing in intensity as it covers a greater area. The area of a sphere being proportioned to the square of the radius, we arrive at the common law that the intensity of sound in still air is inversely proportional to the square of the distance from the source. If in a steady wind the air is moving uniformly at all altitudes, the sound still spreads spherically, but with a moving center,

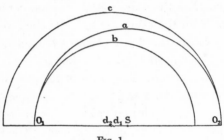

FIG. 1

the whole sphere being carried along. If the air is moving toward the observer, the sound reaches him in less time than it otherwise would, therefore spread over a less spherical surface and louder. If, on the other hand, the observer is to windward, the sound has had to come against the wind, has taken a longer time to reach him, is distributed over a greater surface, and is less loud.

The three cases are represented in the accompanying diagram. The stationary source of sound being at S, a is the wave in still air arriving at both observers at the same time and with the same intensity. If the air is moving to the left, the center of the wave will be shifted by an amount d to the left while the wave has spread to o_1. On arrival it will have the size b, less than a, and will be louder. On the other hand, while the wave is reaching o_2, the observer to windward, the center will have been shifted to the left by an even greater amount d_2. In this case the size of the wave will be c, larger than a, and the sound will be less. The loudness of the sound in the three cases is inversely as the three surfaces a, b, and c. If the distance of the observer from S is denoted by r, the loudness of the sound in the three cases will be as

$$\frac{1}{r^2}, \frac{1}{(r - d_1)^2} \text{ and } \frac{1}{(r + d_2)^2}.$$

The above result may be expressed in the following more simple and practical form. If, in the diagram, a is the wave in still air, its corresponding position when of the same size and, therefore, of the same intensity in moving air will be a', the movement of the air having been sufficient to carry the wave a distance d while it has expanded with the velocity of sound to a sphere of radius r. The distance d and the radius r are to each other as the velocity of wind and the velocity of sound. If the observers o_1 and o_2 move, the one away from the source and the other toward it, by a distance d, the sound will be of the same intensity to both as in their first positions in still air.

In order to make application of this to the particular problem in hand, we shall assume a normal air supply to the room for ventilation purposes of one-seventieth of a cubic meter per person per second. This, if introduced all at one end and exhausted all at the other, in a room 17.9 meters high, 22.8 meters broad, and seating about 2600 persons, would produce a velocity of the air of 0.09 meters per second, assuming the velocity to be the same at every point of a transverse section. Leaving out of account the questionable merits of this arrangement from the ventilation standpoint, its acoustical value can be calculated readily.

The velocity of sound under normal conditions being about 340 meters per second, the time required to traverse a hall 40 meters long is only about one-ninth of a second. In this short interval of time the motion of the air in the room, due to the ventilation, would be sufficient to advance the sound-wave only 0.01 meters, or one centimeter. It would thus arrive at the back of the room as a sphere with its center one centimeter nearer than the source. That is to say, the beneficial effect of this proposed system of ventilation, greatest for the auditor on the rear seat, would to him be equivalent at the very maximum to bringing the stage into the room one centimeter further, or it would be equivalent to bringing the auditor on the rear seat forward one centimeter. This distance is so slight that without moving in his seat, in fact, without moving his shoulders, a slight inclination of the head would accomplish an equivalent gain. Thus, while the effect is in the right direction, it is of entirely imperceptible magnitude. If we take into account the sound reflected from walls and ceiling, the gain is even less.

But the suggestion which is the text of the present paper was not made by one, but by several gentlemen, and is based on the well-recognized fact that one can hear better, often very much better, with the wind than against it, and better than in still air. Therefore, the suggestion is not groundless and cannot be disposed of thus summarily, certainly not without submitting to the same calculation the out-of-door experience that gave rise to the thought.

In the nomenclature of the United States Weather Bureau a wind of from "1 to 5 miles an hour is called light, 6 to 14 miles fresh, 15 to 24 miles brisk, 25 to 37 miles high, and a wind of from 40 to 59 miles is called a gale." Taking the case of a "high wind" as a liberal example, its average velocity is about 14 meters per second, or about one twenty-fifth the velocity of sound. In such a wind the sound 1000 meters to leeward would be louder than in still air only by an amount which would be equivalent to an approach of 40 meters, or 8 per cent. Similarly, to windward the sound would be less loud by an amount equivalent to increasing the distance from 1000 to 1040 meters. This is not at all commensurate with general experience. The difference in audibility, everyone will agree, is generally greater and very much greater than this. The discrepancy, however, can be explained. The discrepancy is not between observation and theory, but between observation and a very incomplete analysis of the conditions in the out-of-door experience. Thus, the ordinary view is that one is merely hearing with or against the wind and this wind is thought of as steady and uniform. As a matter of fact, the wind is rarely steady, and particularly is it of different intensity at different altitudes. Fortunately, the out-of-door phenomenon, which in reality is very complex, has been carefully studied in connection with fog signals.

The first adequate explanation of the variation in loudness of a sound with and against the wind was by the late Sir George G.

Stokes in an article "On the Effect of Wind on the Intensity of Sound," in the *Report of the British Association for the Advancement of Science for 1857*. The complete paper is as follows:

The remarkable diminution in the intensity of sound, which is produced when a strong wind blows in a direction from the observer toward the source of sound, is familiar to everybody, but has not hitherto been explained, so far as the author is aware. At first sight we might be disposed to attribute it merely to the increase in the radius of the sound-wave which reaches the observer. The whole mass of air being supposed to be carried uniformly along, the time which the sound would take to reach the observer, and consequently the radius of the sound-wave would be increased by the wind in the ratio of the velocity of sound to the sum of the velocities of sound and of the wind, and the intensity would be diminished in the inverse duplicate ratio. But the effect is much too great to be attributable to this cause. It would be a strong wind whose velocity was a twenty-fourth part of that of sound; yet even in this case the intensity would be diminished by only about a twelfth part.

The first volume of the *Annales de Chimie* (1816) contains a paper by M. Delaroche, giving the results of some experiments made on this subject. It appeared from the experiments, first, that at small distances the wind has hardly any perceptible effect, the sound being propagated almost equally well in a direction contrary to the wind and in the direction of the wind; second, that the disparity between the intensity propagated in these two directions becomes proportionally greater and greater as the distance increases; third, that sound is propagated rather better in a direction perpendicular to the wind than even in the direction of the wind. The explanation offered by the author of the present communication is as follows:

If we imagine the whole mass of air in the neighborhood of the source of disturbance divided into horizontal strata, these strata do not move with the same velocity. The lower strata are retarded by friction against the earth and by the various obstacles they meet with; the upper by friction against the lower, and so on. Hence, the velocity increases from the ground upward, conformably with observation. This increase of velocity disturbs the spherical form of the sound-wave, tending to make it somewhat of the form of an ellipsoid, the section of which by a vertical diametral plane parallel to the direction of the wind is an ellipse meeting the ground at an obtuse angle on the side towards which the wind is blowing, and an acute angle on the opposite side.

Now, sound tends to propagate itself in a direction perpendicular to the sound-wave; and if a portion of the wave is intercepted by an obstacle of larger size the space behind is left in a sort of sound-shadow, and the only sound there heard is what diverges from the general wave after passing the obstacle. Hence, near the earth, in a direction contrary to the wind, the sound continually tends to be propagated upwards, and consequently there is a continual tendency for an observer in that direction to be left in a sort of sound-shadow. Hence, at a sufficient distance, the sound ought to be very much enfeebled; but near the source of disturbance this cause has not yet had time to operate, and, therefore, the wind produces no sensible effect, except what arises from the augmentation in the radius of the sound-wave, and this is too small to be perceptible.

In the contrary direction, that is, in the direction towards which the wind is blowing, the sound tends to propagate itself downwards, and to be reflected from the surface of the earth; and both the direct and reflected waves contribute to the effect perceived. The two waves assist each other so much the better, as the angle between them is less, and this angle vanishes in a direction perpendicular to the wind. Hence, in the latter direction the sound ought to be propagated a little better than even in the direction of the wind, which agrees with the experiments of M. Delaroche. Thus, the effect is referred to two known causes,— the increased velocity of the air in ascending, and the diffraction of sound.

As a matter of fact, the phenomenon is much more complicated when one takes into consideration the fact that a wind is almost always of very irregular intensities at different altitudes. The phenomenon, in its most complicated form, has been investigated in connection with the subject of fog signals by Professor Osborn Reynolds and Professor Joseph Henry, but with this we are not at present concerned, for the above discussion by Professor Stokes is entirely sufficient for the problem in hand.

The essence of the above explanation is, therefore, this, that the great difference in loudness of sound with and against the wind is not due to the fact that the sound has been simply carried forward or opposed by the wind, but rather to the fact that its direction has been changed and its wave front distorted. The application of this consideration in the present architectural problem leads to the conclusion that the greatest benefit will come not from an attempt to carry the sound by the ventilating movement of the air, but by using the motion of the air to incline the wave front forward and thus direct the sound down upon the audience.

This can be done in either one of two ways, by causing the air to flow through the room from front to back, more strongly at the ceiling than at the floor, or by causing the air to flow from the back to the front, more strongly at the floor than at the ceiling. The one process carrying the upper part of the wave forward, the other retarding the lower part of the wave, will tip the wave in the same way and by an equal amount.

Again, taking an extreme case, the assumption will be made that the motion of the air is such that it is not moving at or near the floor, that it is moving with its maximum velocity at the ceiling, and that the increase in velocity is gradual from floor to ceiling. Keeping the same amount of air moving as in the preceding calculation, the velocity of the air under this arrangement would be twice as great as the average velocity at the ceiling; in the preceding case the wave was advanced one centimeter by the motion of the air while traveling the whole length of the hall. In this case, obviously, the upper part of the wave would be carried twice as far, two centimeters, and the lower part not advanced at all. This would, therefore, measure the total forward tip of the wave.

Fortunately, the acoustical value of this can be expressed in a very simple and practical manner. An inclination of the sound-wave is equivalent acoustically to an equal angular inclination of the floor in the opposite direction. The height of the hall being 17.9 meters, the inclination forward of the sound-wave would be 2 in 1790. The length of the hall being 40 meters, an equal inclination, and thus an equal acoustical effect would be produced by raising the rear of the floor about 5 centimeters. This considers only the sound which has come directly from the stage. It is obvious that if the reflection of the sound from the ceiling and the side walls is taken into account, the gain is even less.

It, therefore, appears that, using the motion of the air in the most advantageous way possible, the resulting improvement in the acoustical property of the hall is of an amount absolutely negligible. A negative result of this sort is perhaps not so interesting as if a positive advantage has been shown; but the problem of properly heating and ventilating a room is sufficiently difficult in itself, and the above considerations are worth while if only to free it from this additional complication.

Temperature

The effect of raising the temperature of a room, involving as it does the contained air and all the reflecting walls and objects, is twofold. It is not difficult to show that, whether we consider the rise in temperature of the air or the rise in temperature of the walls and other reflecting surfaces, the effect of a change of temperature between the limits which an audience can tolerate is negligible, provided the rise in temperature is uniform throughout the room.

The effect of uniformly raising the temperature of the air is to increase the velocity of propagation of sound in all directions. It is, therefore, essentially unlike the effect produced by motion of the air. In the case of a uniform motion of the air, the sound spreads spherically but with unchanged velocity, moving its center in the direction and with the velocity of the wind. Thus, when blown toward the observer, it reaches him as if coming from a nearer source. Blown away from the observer, it arrives as from a more distant source. An increasing temperature of the air increases the velocity, but does not shift the center. The sound reaches the observer coming from a source at an unchanged distance. A rise in temperature, therefore, provided it be uniform, neither increases nor decreases the apparent intensity of the sound. The intensity at all points remains wholly unaltered.

The above is on the assumption that the temperature of the air at all points is the same. If the temperature of the air is irregular, the effect of such irregularity may be pronounced; for example, let us assume a room in which the temperature of the air at the upper levels is greater than at lower levels. In order to make the case as simple as possible, let us assume that the temperature increases uniformly from the floor to the ceiling. To make the case concrete, let us assume that the hall is the same as that described above, practically rectangular, $40 \times 22.8 \times 17.9$ meters. The velocity of the sound at the ceiling, the air being uniform, is greater than it is at the floor. In traversing the room the sound-wave will thus be tipped forward. The effect is practically equivalent as before to an increased pitch of the floor or to an increased elevation of the platform. Without going into the details of this very obvious calculation, it is sufficient to say that in the case of the hall here taken as an example, a difference of temperature top and bottom of 10° C. would be equivalent to an increase in pitch of the floor sufficient to produce an increased elevation of the very back of 10 centimeters. A difference in temperature of 10° C. is not excessive, and it is obvious that this has a greater effect than has that of the motion of the air.

In the above discussion of the effects of motion and of temperature on the acoustical quality of a room, it has been assumed that we are dealing solely with the sound which has come directly from the platform. The argument holds to a less degree for the sound reflected from the ceiling and from the walls. The above estimates, therefore, are outside estimates. The effect is on the whole certainly less. It is safe to say that the total attainable result is not worth the effort that would be involved in altering the architectural features or in compromising the engineering plans.

But, while uniform variation in the motion or in the temperature of the air in the room are on the whole negligible factors in its acoustical character, this is by no means true of irregularities in the temperature of the air, such as would be produced by a column of warm air rising from a floor inlet. That this is a practical point is shown by the testimony of Dr. David B. Reid before the Committee of the Houses of Parliament published in its Report of 1835. This committee was appointed to look into the matter of the heating, ventilation and acoustics of the houses which were being designed to replace those burned in 1832. Of the gentlemen called before the committee, Dr. Reid gave by far the best testimony, part of which was as follows.

Speaking of the hall temporarily occupied by the House of Commons, he said: "Another source of interruption which might be guarded against is the great body of air which I presume arises whenever the heating apparatus is in action below. In different buildings I have had occasion to remark that whenever the atmosphere was preserved in a state of unity as much as possible, equal in every respect, the sound was most distinctly audible; it occurred to me that when the current of hot air rises from the large apparatus in the middle of the House of Commons it would very likely interfere with the communication of sound. On inquiry, one of the gentlemen now present told me he had frequently observed it was impossible to hear individuals who were on the opposite side of this current, although those at a distance were heard distinctly where the current did not intervene." Elsewhere Dr. Reid said: "A current of hot air, rising in a broad sheet along the center of the House, reflected the sound passing from side to side and rendered the intonation indistinct. One of the members of the committee, when I explained this circumstance, stated that he had often noticed that he could not hear a member opposite him distinctly at particular times unless he shifted his seat along the bench, and on examining the place referred to, it was found that he had moved to a position where the hot air current no longer passed between him and the member speaking."

A more recent instance of this sort of difficulty was mentioned to the writer by Mr. W. L. B. Jenney, of Chicago, as occurring in his practice, and later was described in detail in a letter from which the following is quoted:

The building I referred to in my conversation was a court house at Lockport. No plans exist as far as I am aware. Note the sketch I made from remembrance.

Note the passage across the room with stove in center. As the courts were held only during winter there was invariably a fire in that stove. When I examined the room the attendant that was with me informed me that the remarks made by the judge, lawyers and witness could not be heard by the audience on the opposite side of the passageway containing the stove.

At that time, the court room not being occupied, there was no fire in the stove and the doors were closed. I experimented; put the attendant in the judge's stand and took position at "A." I could hear perfectly well. I spoke to him and he replied, "Why, I can hear you perfectly well." I reached this conclusion. That the heated air from the stove and the air supplied by the doors that were constantly fanning at each end of the passageway produced a stratum of air of different density from that of the other parts of the room, which acted like a curtain hanging between the speakers and the hearers. I made my report verbally to the committee that I left below and brought them with me to the room. The experiments were renewed and they accepted my theory. I recommended that the stove be moved and that the warm air should be let into the room from steam coils below at the the end "A" and taken out by exhaust ventilators at the end "B." This was done, and I was informed by the chairman of the committee that the result was very satisfactory. The other conditions of the room were quite usual, — plastering on wooden lath, wooden floors, reasonable height of ceiling.

The above incidents seem to demonstrate fairly clearly that under certain circumstances abrupt irregularities in temperature may result in marked and, in general, unfavorable acoustical effects. The explanation of these effects in both cases is somewhat as follows:

Whenever sound passes from one medium to another of different density, or elasticity, a portion of the sound is reflected. The sound which enters the second medium is refracted. The effects observed above were due to these two phenomena, acting jointly.

The first of the two cases was under simpler conditions, and is, therefore, the easier to discuss. Essentially, it consisted of a large room with speaker and auditor facing each other at a comparatively short distance apart, but with a cylindrical column of hot air rising from a register immediately between. The voice of the speaker, striking this column of air, lost a part by reflection; a part of the sound passed on, entered the column of warm air, and came to the second surface, where a part was again reflected and the remainder went on to the auditor. Thus, the sound in traversing the column of hot air lost by reflection at two surfaces and reached the auditor diminished in intensity. It reached the auditor with diminished intensity for another reason.

The column of warm air acted like a lens. The effect of the column of air was not like that of the ordinary convex lens, which would bring the sound to a focus, but rather as a diverging lens. The effect of a convex lens would have been obtained had the column of air been colder than that of the surrounding room. Because the air was warmer, and, therefore, the velocity of sound through it greater, the effect was to cause the sound in passing through the cylindrical column to diverge even more rapidly and to reach the auditor very considerably diminished in intensity. Which of these two effects was the more potent in diminishing the sound, whether the loss by reflection or the loss by lens-like dispersion was the greater, could only be determined if one knew the temperature of the air in the room, in the column, and the diameter of the column. It is sufficient, perhaps, to point out on the authority of such eminent men as Dr. Reid and Mr. Jenney that the phenomenon is a real one and one to be avoided, and that the explanation is ready at hand and comparatively simple.

It is, perhaps, worth while pointing out that in both of the above cases there was a good deal of reverberation in the room, so that any considerable diminution in the intensity of the sound coming directly from the speaker to the auditor resulted in its being lost in the general reverberation. Had the same conditions as to location of speaker, auditor, column of warm air and temperature occurred out of doors or in a room of very slight reverberation the effect would have been very much less noticeable. Nevertheless, great irregularity of temperature is to be avoided, as the above testimony fairly clearly shows.

The above also suggests another line of thought. If, instead of having a single screen of great temperature difference between speaker and auditor, there were many such differences in temperature, though slight in amount, the total effect might be great. This corresponds, in the effect produced, to what Tyndall calls a "flocculent condition of the atmosphere" in his discussion of the transmission of fog signals. Tyndall points out that if the atmosphere is in layers alternately warm and cold sound is transmitted with much more rapid diminution in intensity than when the atmosphere is of very uniform temperature. This phenomenon is, of course, much more important with such temperature differences as occur out of doors than in a room, but it suggests that, in so far as it is a perceptible effect, the temperature of a room should be homogeneous. This condition of homogeneity is best secured by that system of ventilation known as "distributed floor outlets." It has the additional merit of being, perhaps, the most efficient system of ventilation.

5

SENSE OF LOUDNESS [1]

IT will be shown here that there is a sense of relative loudness, particularly of equality of loudness, of sounds differing greatly in pitch, that this sense of loudness is accurate, that it is nearly the same for all normal ears, that it is independent of experience, and that, therefore, it probably has a physical and physiological basis. This investigation has been incidental to a larger investigation on the subject of architectural acoustics. It has bearing, however, on many other problems, such, for example, as the standardization of noises, and on the physiological theory of audition.

The apparatus used consisted of four small organs (Proc. Am. Acad. of Arts and Sciences, 1906)[2] so widely separated from each other as to be beyond the range of each other's influence. Each organ carried seven night-horn organ pipes at octave intervals in pitch, 64, 128, 256, . . . 4096 vibrations per second. The four organs were so connected electrically to a small console of seven keys that on pressing one key, any one, any two, any three or all four organ pipes of the same pitch would sound at once, — the combination of organ pipes sounding being adjusted by an assistant and unknown to the observer.

In other parts of the investigation on architectural acoustics the loudness of the sound emitted by each of the twenty-eight organ pipes in terms of the minimum audible sound for the corresponding pitch had been determined. The experiment was conducted in the large lecture-room of the Jefferson Physical Laboratory, and, in the manner explained elsewhere, the computation was made for the loudness of the sound, taking into account the shape of the room and the materials employed in its construction.

The experiment consisted in adjusting the number of pipes which were sounding or in choosing from among the pipes until such an adjustment was accomplished, that, to an observer in a more or less remote part of the room all seven notes, when sounded in succession, seemed to have the same loudness. As the pipes of the same pitch

[1] Contributions from the Jefferson Physical Laboratory, vol. viii, 1910.
[2] See p. 84.

did not all have the same loudness, it was possible by taking various combinations to make this adjustment with considerable accuracy. This statement, however, is subject to an amendment in that all four pipes of the lowest pitch were not sufficiently loud and the faintest of the highest pitch was too loud.

There were ten observers, and each observer carried out four independent experiments. Speaking broadly, in the case of every observer, the four independent experiments agreed among themselves with great accuracy. This was to the great surprise of every observer, each before the trial doubting the possibility of such adjustment. The results of all ten observers were surprisingly concordant.

After the experiment with the first two observers, it seemed possible that their very close agreement arose from their familiarity with the piano, and that it might be that they were adjusting the notes to the "balance" of that particular instrument. The next observer, therefore, was a violinist. Among the observers there was also a 'cellist. Lest the feeling of relative loudness should come from some subconscious feeling of vocal effort, although it is difficult to see how this could extend over so great a range as six octaves, singers were tried whose voices were of very different register. Two of the observers, including one of the pianists, were women. Two of the observers were non-musical, one exceedingly so.

The accompanying table gives the results of the observations, the energy of each sound being expressed in terms of minimum audible intensity for that particular pitch, after making all corrections for the reënforcement of the sound by the walls of the room. The observations are recorded in order, the musical characteristic of the observer being indicated.

PITCH FREQUENCY

Observers	64	128	256	512	1024	2048	4096
1. Piano	$7.0(+)\times10^4$	1.7×10^6	4.4×10^6	8.0×10^6	15.0×10^6	9.6×10^5	$4.5(-)\times10^3$
2. Piano	7.0+	1.7	4.4	11.2	9.2	12.0	5.2—
3. Non-musical	7.0+	1.7	3.6	8.9	6.3	9.6	4.5—
4. Non-musical	7.0+	1.7	3.7	7.7	14.5	14.4	5.6—
5. Violin	7.0+	1.7	3.5	11.7	13.9	8.0	3.5—
6. Violin	7.0+	1.7	4.0	11.4	15.5	15.2	5.2—
7. 'Cello	7.0+	1.7	4.2	12.0	13.4	9.6	5.1—
8. Tenor	7.0+	1.7	3.9	13.3	13.5	10.5	4.0—
9. Soprano	7.0+	1.7	4.7	12.9	17.0	9.6	5.4—
10. Piano	7.0+	1.7	3.5	13.2	14.5	8.0	4.9—
	7.0(+)	1.7	4.0	11.0	13.3	10.6	4.8—

ARCHITECTURAL ACOUSTICS[1]

CORRECTION OF ACOUSTICAL DIFFICULTIES

ON the completion of the Fogg Art Museum in 1895, I was requested by the Corporation of Harvard University to investigate the subject of architectural acoustics with the end in view of correcting the lecture-room which had been found impracticable and abandoned as unusable. Later the planning of a new home for the Boston Symphony Orchestra in Boston widened the scope of the inquiry. Since then, over questions raised first by one building and then another, the subject has been under constant investigation.

In 1900 a series of articles, embodying the work of the first five years and dealing with the subject of reverberation, was published in the American Architect and also in the Engineering Record. The next five years were devoted to the extension of this study over the range of the musical scale and the results were published in the Proceedings of the American Academy of Arts and Sciences in 1906. Since then the investigation has been with reference to interference and resonance, the effects of peculiarities of form, and the causes of variation in audibility in different parts of an auditorium. These results will be published in another article during the ensuing year.

The progress of this experimental investigation has been guided in practical channels and greatly enriched by the experience gained from frequent consultation by architects, either for purposes of correcting completed buildings or in the preparation of plans in advance of construction. Reserving for a later article the stimulating subject of advance planning, the present article is devoted to the problems involved in the correction of completed buildings. It is illustrated by a few examples which are especially typical. I desire to take this opportunity of expressing my appreciation of the very cordial permission to use this material given by the architects, Messrs. McKim, Mead & White, Messrs. Carrère & Hastings, Messrs. Cram, Goodhue & Ferguson, and Messrs. Allen & Collens

[1] The Architectural Quarterly of Harvard University, March, 1912.

— to these and to the other architects whose confidence in this work has rendered an extensive experience possible.

The practical execution of this work of correction has recently been placed on a firmer basis by Mr. C. M. Swan, a former graduate student in the University and an associate in this work, who has taken charge of a department in the H. W. Johns-Manville Company. I am under obligations to him and to this company for some of the illustrations used below, and to the company, not merely for having placed at my disposal their materials and technical experience, but also for having borne the expense of some recent investigations looking toward the development of improved materials, with entire privilege of my making free publication of scientific results.

It is proposed to discuss here only such corrective methods as can be employed without extensive alterations in form. It is not proposed to discuss changes of dimension, changes in the position of the wall-surfaces or changes in ceiling height. It is the purpose to discuss here medicinal rather than surgical methods. Such treatment properly planned and executed, while not always available, will in the great majority of cases result in an entire remedy of the difficulty.

Two old, but now nearly abandoned devices for remedying acoustical difficulties are stretched wires and sounding-boards. The first is without value, the second is of some value, generally slight, though occasionally a perceptible factor in the final result. The stretching of wires is a method which has long been employed, and its disfiguring relics in many churches and court rooms proclaim a difficulty which they are powerless to relieve. Like many other traditions, it has been abandoned but slowly. The fact that it was wholly without either foundation of reason or defense of argument made it difficult to answer or to meet. The device, devoid on the one hand of scientific foundation, and on the other of successful experience, has taken varied forms in its application. Apparently it is a matter of no moment where the wires are stretched or in what amount. There are theatres and churches in Boston and New York in which four or five wires are stretched across the middle of the room; in other auditoriums miles on miles of wire have been stretched; in both it is equally without effect. In no case can one obtain more than a qualified approval, and the most earnest negatives come where the wires have been used in the largest amount. Occasionally the response to inquiries is that "the wires may have done some good but certainly not much," and in general when even that qualified approval is given the installation of the wires was

FIG. 1. Ceiling of church, San José, California, showing an ineffective use of stretched wires.

accompanied by some other changes of form or occupancy to which the credit should be given. How extensive an endeavor is sometimes made in the use of stretched wires is shown by the accompanying illustration which shows a small section of the ceiling of a church in San José, California. In this church between one and two miles of wire have been stretched with resulting disfigurement, and wholly without avail. The question is being taken up again by the church for renewed effort.

Aside from such cumulative evidence of ineffectiveness, it is not difficult to show that there is no physical basis for the device. The sound, whose echoes these wires are presumed to absorb, scarcely affects the wires, giving to them a vibration which at most is of microscopical magnitude. If the string of a violin were free from the body of the violin, if the string of a piano were free from the

Fig. 2. Congregational Church, Naugatuck, Connecticut. McKim, Mead and White, Architec

sounding-board, if the string of a harp did not touch the thin sounding-board which faces its slender back, when plucked they would not emit a sound which could be heard four feet away. The sound which comes from each of these instruments is communicated to the air by the vibration of its special sounding-board. The string itself cuts through the air with but the slightest communication of motion. Conversely when the sound is in the room and the string at rest the vibrating air flows past it, to and fro, without disturbing it, and consequently without itself being affected by reaction either for better or worse.

The sounding-board as a device for correcting acoustical difficulties has at times a value; but unless the sounding-board is to be a large one, the benefit to be expected from its installation may be greatly overrated. As this particular subject calls for a line of

FIG. 3. Hall of the House of Representatives, Rhode Island State Capitol, Providence, R. I. McKim, Mead and White, Architects.

argument very different from that of the main body of the present paper, it will be reserved for a discussion elsewhere, where, space permitting, it can be illustrated by examples of various forms accompanied by photographs and by a more or less exhaustive discussion of their relative merits.

The auditorium in whose special behalf this investigation started seventeen years ago was the lecture-room of the Fogg Art Museum.

Although this room was in a large measure remedied, it will not be taken as an example. Its peculiarities of shape were such that its complete relief was inherently a complicated process. While this case was chronologically the first, it is thus not suitable for an opening illustration.

Among a number of interesting problems in advance of construction the firm of McKim, Mead & White has brought some

Fig. 4.   Detail, Hall of the House of Representatives, Rhode Island State Capitol.
McKim, Mead and White, Architects.

interesting problems in correction, of which three will serve admirably as examples because of their unusual directness. The first is that of the Congregational Church in Naugatuck, Connecticut, shown in the accompanying illustration. When built its ceiling was cylindrical, as now, but smooth. Its curvature was such as to focus a voice from the platform upon the audience, — not at a point, but along a focal line, for a cylindrical mirror is astigmatic. The difficulty was evident with the speaking, but may be described more effectually with reference to the singing. The position of the choir was behind the preacher and across the main axis of the church. On one line in the audience, crossing the church obliquely from right to left, the soprano voice could be heard coming even more sharply from the ceiling than directly from the singer. The alto starting nearer the axis of the church had for its focus a line crossing the church less obliquely. The phenomena were similar for the tenor and the bass voices, but with focal lines crossing the church obliquely in opposite directions. The difficulty was in a very large measure remedied by coffering the ceiling, as shown in the illustration, both the old and the new ceiling being of plaster. Ideally a larger and deeper coffering was desirable, but the solution as shown was practical and the result satisfactory.

The hall of the House of Representatives in the Rhode Island State Capitol illustrated another type of difficulty. In considering this hall it is necessary to bear in mind that the problem is an essentially different one from that of a church or lecture-room. In these the speaking is from a raised platform and a fixed position. In a legislative assembly the speaking is in the main from the floor, and may be from any part of the floor; the speaker stands on a level with his fellow members; he stands with his back to a part of the audience, and often with his back to the greater part of his audience; in different parts of the house the speaker directs his voice in different directions, and against different wall-surfaces. In this hall the walls were of stone to approximately half the height of the room; above that they were of stone and plaster. The ceiling was, as shown, coffered. The difficulty in this room was with that part of the voice which, crossing the room transversely, fell on the side walls. With the speaker standing on the floor, the greater volume of his voice was directed upward. The sound striking the side wall was reflected across the room to the opposite wall and back again, to and fro, mounting gradually until it reached the ceiling. It was there reflected directly down upon the audience. The ceiling sloped, and had some curvature, but the curvature was not such as to produce a distinct focusing of the sound. During these reflections the sound met only feebly absorbent surfaces and therefore returned to the audience with but little loss of intensity. Its return was at such an interval of time as to result in great confusion of speech. Only the fact that the voice, rising at different angles, traveled different paths and therefore returned at varying intervals, prevented the formation of a distinct echo. The difficulty was remedied in this case by a change in material without change

Fig. 5. Lecture-room, Metropolitan Museum of Art, New York.
McKim, Mead and White, Architects.

of form, by diminishing the reflecting power of the two side walls. This was done by placing a suitable felt on the plaster walls between the engaged columns, and covering it with a decorated tapestry. Fortunately, the design of the room admitted of a charming execution of this treatment. It is interesting to note that this treatment applied to the lower half of the walls would not have been acoustically effective.

The lecture-room of the Metropolitan Museum of Art illustrates the next step in complexity. This hall is a semi-circular auditorium, with the semi-circle slightly continued by short, straight walls. As shown in the illustrations the platform is nearly, though not wholly, within a broad but shallow recess. The body

FIG. 6. Lecture-room, Metropolitan Museum of Art, New York. McKim, Mead and White, Architects.

of the auditorium is surmounted by a spherical ceiling with short cylindrical extension following the straight side walls. In the center of the ceiling is a flat skylight of glass. In this room the reverberation was not merely excessive, but it resolved itself by focusing into a multiple echo, the components of which followed each other with great rapidity but were distinctly separable. The number distinguishable varied in different parts of the hall. Seven were distinguishable at certain parts. A detailed discussion of this is not appropriate in the present paper as it concerns rather the subject of calculation in advance of construction. To improve the acoustics the ceiling was coffered, the limiting depth and dimensions of this coffering being determined in large measure by the dimensions of the skylight. The semi-circular wall at the rear of the auditorium was transformed into panels which were filled with felt over which was stretched burlap as shown in the second illustration. The result was the result assured,—the reduction of the disturbance to a single and highly localized echo. This echo is audible only in the central seats — two or three seats at a time — and moves about as the speaker moves, but in symmetrically opposite direction. Despite this residual effect, and it should be noted that this residual effect was predicted, the result is highly satisfactory to Dr. Edward Robinson, the Director of the Museum, and the room is now used with comfort, whereas it had been for a year abandoned.

It should be borne in mind that "perfect acoustics" does not mean the total elimination of reverberation, even were that possible. Loudness and reverberation are almost, though not quite, proportional qualities. The result to be sought is a balance between the two qualities, dependent on the size of the auditorium and the use to which it is to be applied.

Geometrically the foregoing cases are comparatively simple. In each case the room is a simple space bounded by plane, cylindrical or spherical surfaces, and these surfaces simply arranged with reference to each other. The simplicity of these cases is obvious. The complexity of other cases is not always patent, or when patent it is not obvious to a merely casual inspection how best the problem should be attacked. A large number of cases, however, may be handled in a practical manner by regarding them as connecting spaces, each with its own reverberation and pouring sound into and receiving sound from the others. An obvious case of this is the theatre, where the aggregate acoustical property is dependent on the space behind the proscenium arch in which the speaker stands, as well as on the space in front of it. In another sense and to a less degree, the cathedral, with its chancel, transept and nave may be

FIG. 7. Design for St. Paul's Cathedral, Detroit. Cram, Goodhue and Ferguson, Architects.

regarded as a case of connected spaces. The problem certainly takes on a simpler aspect when so attacked. An extreme and purely hypothetical case would be a deep and wide auditorium with a very low ceiling, and with a stage recess deep, high and reverberant, in fact such a case as might occur when for special purposes two very different rooms are thrown together. In such a case the reverberation calculated on the basis of a single room of the combined volume and the combined absorbing power would yield an erroneous value. The speaker's voice, especially if he stood back some distance from the opening between the two rooms, would be lost in the production of reverberation in its own space. The total resulting sound, in a confused mass, would be propagated out over the auditorium. Of course this is an extreme case and of unusual occurrence, but by its very exaggeration serves to illustrate the point. In a less degree it is not of infrequent occurrence. It was for this reason, or rather through the experience of this effect, although only as a nice refinement, that the Boston Symphony Orchestra has its special scenery stage in Carnegie Hall, and for this that Mr. Damrosch in addition moved his orchestra some little distance forward into the main auditorium for his concerts in the New Theatre.

A cathedral is a good example of such geometrical complication, still further complicated by the variety of service which it is to render. It must be adapted to speaking from the pulpit and to reading from the lectern. It must be adapted to organ and vocal music, and occasionally to other forms of service, though generally of so minor importance as to be beyond the range of appropriate consideration. Most cathedrals and modern large churches have a reverberation which is excessive not only for the spoken but also for a large portion of the musical service. The difficulty is not peculiar to any one type of architecture. To take European examples, it occurs in the Classic St. Paul in London, the Romanesque Durham, the Basilican Romanesque Pisa, the Italian Gothic Florence, and the English Gothic York.

The most interesting example of this type has been Messrs. Cram, Goodhue & Ferguson's charming cathedral in Detroit, especially interesting because in the process of correcting the acoustics it was possible to carry to completion the decoration of the original design.

Fig. 8. St. Paul's Cathedral, Detroit. Cram, Goodhue and Ferguson, Architects.

The nave, moderately narrow in the clerestory, was broad below through its extension by side aisles. It might fairly be regarded as two simply connected spaces. The lower space, when there was a full audience, was abundantly absorbent; the clerestory, though with wood ceiling, was not absorbent. Although their combined reverberation was great, it was not so great as alone to produce the actual effect obtained. Absorbing material in the form of a felt, highly efficient acoustically, was placed in the panels on the ceiling. The original architectural design by Mr. Cram (Fig. 7) showed the ceiling decorated in colors, and this though not a part of the original construction was carried out on the covering of the felt, with a result highly satisfactory both acoustically and architecturally. The transept, also high and reverberant, was similarly treated, as was also the central tower which was even higher than the rest of the church. As a matter of fact the results at first attained were satisfactory only with an audience filling at least three-quarters of the seats, the condition for which it was planned. The treatment was subsequently extended to the lower levels in order that the cathedral might be serviceable not merely for the normal but for the occasionally small audience. The chancel did not need and did not receive any special treatment. It was highly suitable to the musical service, and being at the back of both the pulpit and the lectern did not greatly affect that portion of the service which called for distinctness of enunciation.

It may be remarked in passing that the lectern is almost invariably a more difficult problem than the pulpit. This is in part because reading, with the head thrown slightly forward, is more difficult than speaking; because, if the lectern is sufficiently high to permit of an erect position it screens the voice; because a speaker without book or manuscript, seeing his audience, realizes his distance and his difficulties; and finally, because the pulpit is generally higher and against a column whereas the lectern stands out free and unsupported.

The auditorium which has received the greatest amount of discussion recently is the New Theatre in New York. Had it been a commercial proposition its acoustical quality would have received but passing notice. As an institution of large purpose on the part of the Founders it received a correspondingly large attention. As an institution of generous purpose, without hope or desire for financial return, it was appropriated by the public, and received the persistent criticism which seems the usual reward for such undertakings. The writer was consulted only after the completion of the building, but its acoustical difficulties can be discussed adequately only in the light of its initial programme.

It was part of the original programme submitted to Messrs. Carrère & Hastings that the building should be used, or at least should be adapted to use for opera as well as for drama. In this respect it was to bear to the Metropolitan the position which the Opera Comique in Paris bears to the Opera. This idea, with its corollary features, influenced the early design and shows in the completed structure.

It was also a part of the initial plan that there should be two rows of boxes, something very unusual in theatre construction. This was a prodigal use of space and magnified the building in all its dimensions. Later, but not until after the building was nearly completed, the upper row of boxes was abandoned, and the gallery thus created was devoted to foyer chairs. As the main walls were by this time erected, the gallery was limited in depth to the boxes and their antechambers. It thus resulted that this level, which is ordinarily occupied by a gallery of great value, is of small capacity. Notwithstanding this the New Theatre seats twenty-three hundred, while the usual theatre seats but little more than two-thirds that number.

The necessity of providing twenty-three commodious boxes, all in the first tier, of which none should be so near the stage as to be distinctly inferior, determined a large circle for their front and for the front of all the galleries. Thus not merely are the seats, which are ordinarily the best seats, far from the stage, but the great horizontal scale thus necessitated leads architecturally to a correspondingly great vertical scale. The row of boxes and the foyer balcony above not merely determined the scale of the auditorium, but also presented at the back of their shallow depth a concave wall which focused the reflected sound in the center of the auditorium.

Finally, it should be borne in mind that while the acoustical demands in a theatre are greater than in almost any other type of auditorium, because of the great modulation of the voice in dramatic action, the New Theatre was undertaking an even more than usually difficult task, that of presenting on the one hand the older dramas with their less familiar and more difficult phrasing, and on the other the more subtle and delicate of modern plays.

FIG. 9.   Interior, the New Theatre, New York City.   Carrère and Hastings, Architects.

The conventional type of theatre construction is fairly, though only fairly, well adapted to the usual type of dramatic performance. The New Theatre, with a very difficult type of performance to present, was forced by the conditions which surrounded the project to depart from the conventional type far more radically than was perhaps at that time realized.

Here, as usual in a completed building, structural changes and large changes of form were impossible, and the acoustical difficulties of the auditorium could be remedied only by indirection. The method by which a very considerable improvement was attained is shown by a comparison of the line drawing (Fig. 10) with the photograph of the interior of the theatre as originally completed. The boxes were changed from the first to the second level, being interchanged with the foyer chairs, while the excessive height of the main body of the auditorium was reduced by means of a canopy surrounding the central chandelier. This ingenious and not dis-

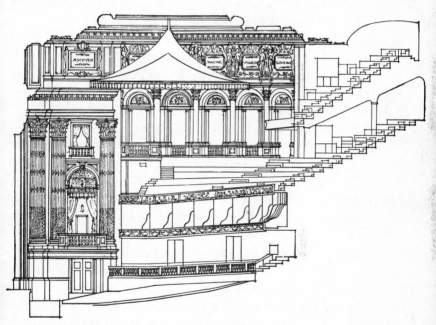

FIG. 10. The New Theatre, New York City, showing Canopy and Changed Boxes.

pleasing substitute for the recommended lowering of the ceiling was proposed by Mr. Hastings, although of course only as a means to an end. The canopy is oval in plan, following the outline of the oval panel in the ceiling, its longer axis being transverse. Its major and minor horizontal dimensions are 70 feet and 40 feet. Its effective lowering of the height of the ceiling is 20 feet. A moment's consideration will show that its effective area in preventing the ceiling echo is greater than its actual dimensions, particularly in the direction of its minor axis. The improvement brought about by this was pronounced and satisfactory to the Founders. The distances, however, were still too great, even visually, for the type of dramatic performance for which the theatre was primarily intended, and such use was therefore discontinued. The New Theatre is much better adapted to opera than to dramatic performances, and it will be a matter of great regret if, with its charming solution of many difficult architectural problems, it is not restored to such dignified purpose.

The last and very satisfactory example is that of the Chapel of the Union Theological Seminary of Messrs. Allen & Collens. Its interesting feature is that the corrective treatment was applied in the process of construction. It is further interesting as an example of a treatment which is not merely inconspicuous, but is entirely indistinguishable. The photograph without explanation is the best evidence of this (p. 149).

The above examples have been chosen from many score as typical of the principles involved. In each case the nature of the difficulty has been stated and the method employed in its correction, or at least its special feature very briefly described. The remainder of the paper will be devoted to a discussion of the principles involved in acoustical correction and in presenting the results of some recent experiments.

In discussing the above examples, especially the first and the third, the Congregational Church in Naugatuck, and the lecture-room of the Metropolitan Museum of Art, consideration had to be given to the effect of the geometrical shape of the room. This aspect of the problem of architectural acoustics constitutes a subject so large that a separate paper must be devoted to its adequate treatment. It involves not merely simple reflection but interference and diffraction, as well as the far from simple subject of the propagation of sound parallel to or nearly parallel to the plane of an audience. It has been the object of special investigation during the past six years. This investigation has recently come to a successful issue and will probably be published in full during the ensuing year. It is suitable that it should receive separate publication for, as it concerns shape, it is of more value for calculation in ad-

Fig. 11. Chapel, Union Theological Seminary, New York City. Allen and Collens, Architects.

vance of construction than in the correction of completed buildings. It must here suffice to merely indicate the nature of the results.

When sound is produced in a confined auditorium it spreads spherically from the source until it reaches the audience, the walls, or the ceiling. It is there in part absorbed and in part reflected. The part which is reflected retraverses the room until it meets another surface. It is again in part absorbed and in part reflected. This process continues until, after a greater or less number of reflections, the sound becomes of negligible intensity. Thus at any one time and at any one point in the room there are many sounds crossing each other. In a very simple auditorium, such as a simple rectangular room with plain walls and ceiling, this process is not difficult to follow, either step by step, or by large, but entirely adequate, generalizations. When the conditions are more complicated it is more difficult to analyze; it is also more liable to be a vitally significant factor in the problem. That it has heretofore been inadequately discussed has arisen from the failure to take into consideration the phenomenon of diffraction in the propagation of a sound nearly parallel to an absorbing audience, the phenomenon of diffraction in reflection from an irregular surface, and, above all, the phenomenon of interference. The first of these three considerations is of primary importance in calculating the intensity of the sound which has come directly from the source, in calculating the effect of distance in the audience, and in calculating the relative loudness on the floor and in the gallery, and at the front and at the back of the gallery. The second consideration enters into the calculation of the path of the sound after reflection from any broken or irregular surfaces. The third is a factor of the utmost importance when the sounds which are crossing at any point in the auditorium are of comparable intensity and have traveled paths of so nearly equal length that they have originated from the same element. This latter calls for a more elaborate explanation.

In both articulate speech and in music the source of sound is rapidly and in general, abruptly changing in pitch, quality, and loudness. In music one pitch is held during the length of a note. In articulate speech the unit or element of constancy is the syllable. Indeed, in speech it is even less than the length of a syllable, for the open vowel sound which forms the body of a syllable usually has a consonantal opening and closing. During the constancy of an element, either of music or of speech, a train of sound-waves spreads spherically from the source, just as a train of circular waves spreads outward from a rocking boat on the surface of still water. Different portions of this train of spherical waves strike different surfaces of the auditorium and are reflected. After such reflection they begin to cross each other's paths. If their paths are so different in length that one train of waves has entirely passed before the other arrives at a particular point, the only phenomenon at that point is prolongation of the sound. If the space between the two trains of waves be sufficiently great the effect will be that of an echo. If there be a number of such trains of waves thus widely spaced, the effect will be that of multiple echoes. On the other hand if the two trains of waves have traveled so nearly equal paths that they overlap, they will, dependent on the difference in length of the paths which they had traveled, either reënforce or mutually destroy each other. Just as two equal trains of water-waves crossing each other may entirely neutralize each other if the crest of one and the trough of the other arrive together, so two sounds, coming from the same source in crossing each other may produce silence. This phenomenon is called interference and is a common phenomenon in all types of wave motion. Of course this phenomenon has its complement. If the two trains of water-waves so cross that the crest of one coincides with the crest of the other and trough with trough, the effects will be added together. If the two sound-waves be similarly retarded, the one on the other, their effects will also be added. If the two trains of waves be equal in intensity, the combined intensity will be quadruple that of either of the trains separately, as above explained, or zero, depending on their relative retardation. The effect of this phenomenon is to produce regions in an auditorium of loudness and regions of comparative or even complete silence. It is a partial explanation of the so-called deaf regions in an auditorium.

It is not difficult to observe this phenomenon directly. It is difficult, however, to measure and record the phenomenon in such a manner as to permit of an accurate chart of the result. Without going into the details of the method employed the result of these measurements for a room very similar to the Congregational Church in Naugatuck is shown in the accompanying chart. The room experimented in was a simple rectangular room with plain side

FIG. 12. Distribution of intensity on the head level in a room with a barrel-shaped ceiling, with center of curvature on the floor level.

walls and ends and with a barrel or cylindrical ceiling. The ceiling of the room was smooth like the ceiling of the Naugatuck Church before it was coffered. The result is clearly represented in Fig. 12, in which the intensity of the sound has been indicated by contour lines in the manner employed in the drawing of the Geodetic Survey maps. The phenomenon indicated in these diagrams was not ephemeral, but was constant so long as the source of sound continued, and repeated itself with almost perfect accuracy day after day. Nor was the phenomenon one which could be observed merely instrumentally. To an observer moving about in the room it was quite as striking a phenomenon as the diagrams suggest. At the points in the room indicated as high maxima of intensity in the diagram the sound was so loud as to be disagreeable, at other points so low as to be scarcely audible. It should be added that this distribution of intensity is with the source of sound at the center of the room. Had the source of sound been at one end and on the axis of the cylindrical ceiling, the distribution of intensity would still have been bilaterally symmetrical, but not symmetrical about the transverse axis.

As before stated a full discussion of this phase of the subject is reserved for another paper which is now about ready for publication.

In the second, in the fourth, and in part in the third of the above examples the acoustical difficulty was that of excessive reverberation.

If a sound of constant pitch is maintained in an auditorium, though only for a very brief time, the sound spreading directly from the source, together with the sound which has been reflected, arrives at a steady state. The intensity of the sound at any one point in the room is then the resultant of all the superposed sounds crossing at that point. As just shown, the mutual interference of these superposed sounds gives a distribution of intensity which shows pronounced maxima and minima. However, the probable intensity at any point, as well as the aggregate intensity over the room, is the sum of the components. Whatever the distribution of maxima and minima the state is a steady one so long as the source continues to sound. The steady condition in the room is such that the rate of absorption of the sound is equal to the rate of emission by the source.

If after this steady state is established the source is abruptly checked, the different trains of waves will continue their journey, the maxima and minima shifting positions. Ultimately, the sound will cease to be audible, having diminished in intensity until it has passed below what aurists call the "threshold of audibility." The duration of audibility after the source has ceased is thus dependent upon the initial intensity, upon the absorbing material, and upon the location of that absorbing material with reference to the several trains of waves. In special cases the position of the absorbing material is a matter of the utmost importance, but in many cases the aggregate result may be computed on the basis of the total absorbing power in the room.

The prolongation of the sound in an auditorium after the source has ceased I have ventured to call reverberation, and to measure it numerically by the duration of audibility after the abrupt cessation of a source which has produced an average intensity of sound in the room equal to one million times minimum audible intensity. This is an ordinary condition in actual occurrence.

In the 1900 papers published in the Engineering Record and the American Architect, this subject of reverberation was discussed at great length, and it was there shown how it might be measured and indeed, how it might be calculated in advance of construction. In addition to the formula many coefficients of absorption were determined, such data being absolutely necessary to the reduction of the subject to an exact science. This work related to sounds having a pitch an octave above middle C.

But it was of course obvious that the acoustical quality of an auditorium is not determined by its character with reference to a single note. The next series of papers, published in 1906, therefore extended the investigation over the whole range of the musical scale giving data for many materials and wall-surfaces, and rendering a more complete calculation possible. At the conclusion of these papers it was shown how the reverberation of an auditorium should be represented by a curve in which the reverberation is plotted against the pitch and by way of illustration a particular case was shown, that of the large lecture-room in the Jefferson Physical Laboratory, both with and without an audience. This curve is reproduced in the accompanying diagram (Fig. 13).

In the process of investigating an auditorium such a curve should be drawn as definitive of its initial condition and then in the determination of the treatment to be employed similar curves should be drawn representing the various alterations proposed and taking into consideration the location of the surfaces, their areas and the nature of the proposed treatment. The diagram (Fig. 14) shows the result of this computation for the more interesting of the above examples, St. Paul's Cathedral, Detroit. In this diagram curves are drawn plotting the reverberation of the

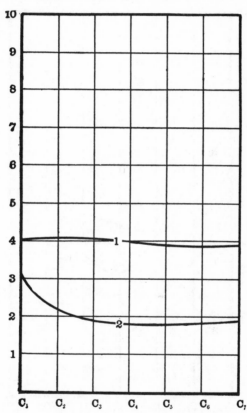

Fig. 13. Curves showing the reverberation in the lecture-room of the Jefferson Physical Laboratory without an audience and with an audience filling all the seats.

cathedral in its original condition, empty, and with a three-quarters audience, and with a full audience, and again after its acoustical correction also empty, with a three-quarters audience, and with a full audience.

Reprints of the papers just mentioned were mailed at the time to all members of the American Institute of Architects. Duplicates will gladly be sent to any one who may be interested in the further perusal of the subject.

Brief mention has been made of the dependence, in special cases, of the efficiency of an absorbing material on its positions in an auditorium. For example, in the room whose distribution of intensity

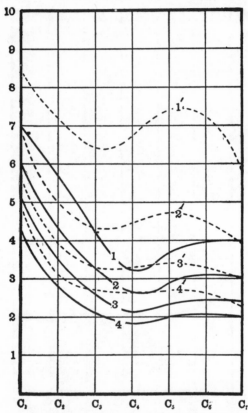

Fig. 14. Curves showing the reverberation in St. Paul's Cathedral, Detroit, before (1′, 2′, 3′, 4′) and after (1, 2, 3, 4) corrections, empty and with a one-quarter, one-half, three-quarter and full audience.

was shown in Fig. 12, the absorbing material would have much greater efficiency in reducing the reverberation if placed so as to include maxima, than if so placed as to include minima. That this would be true is obvious. The magnitude of the effect, however, is not so clear, for the maxima and minima shift as the sound dies away. It was therefore submitted to an accurate experimental investigation. The results are shown in the adjacent diagram,

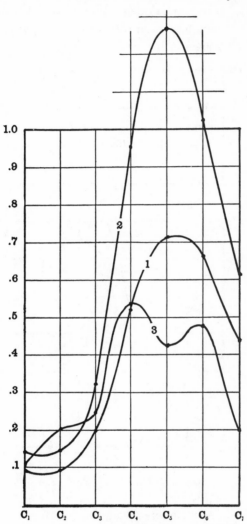

FIG. 15. Showing the relative efficiency of felt in different parts of a room having a barrel ceiling. Curve 1, normal absorbing power; Curve 2, absorbing power in the center of the room; Curve 3, absorbing power at the side of the room. C_3 is middle C, 256.

Fig. 15. In this diagram the curve marked 1 shows by its vertical ordinates the normal efficiency of a very highly absorbent felt. If so placed in the room as to include on its surface the maxima of intensity of the sound it had an effective absorbing power as shown in Curve 2, a truly remarkable increase over its normal value. Curve 3 shows the efficiency of the same felt when placed against the side wall. It there included more maxima than minima for the

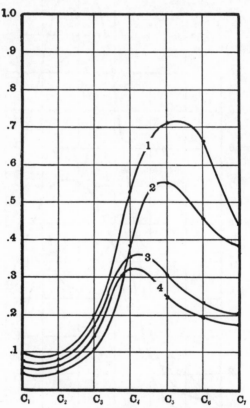

FIG. 16. Absorbing power of various kinds of felt as defined in the text. C_3 is middle C, 256.

lower notes, but more minima than maxima for the higher notes, with a resulting efficiency curve which is very irregular.

The following experiments were performed for the H. W. Johns-Manville Company in the search for an efficient absorbing material and an effective method of treatment. The absorbing efficiency of felt is dependent on the flexibility of the mass as a whole and on its porosity. It is not in large measure dependent on the material employed, except in so far as the nature of that material determines the nature, and therefore the closeness, of the felting process. The same materials, therefore, might very well have either a very high or a very low absorbing efficiency, depending entirely upon the process of manufacture. The nature of the material is here specified,

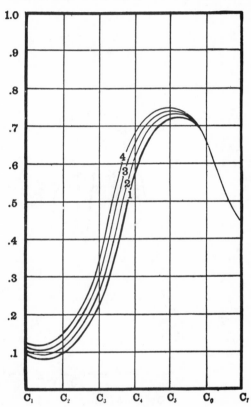

Fig. 17. Effect of air space behind felt. Curve 1, felt in contact with the wall; Curves 2, 3, and 4, felt at distances of 2, 4, and 6 inches from the wall.

not with the idea that it alone can determine the quality, but merely as an additional piece of information. In addition to this, in each case the ratio of the solid material to the free space is given; but even this does not define in full the essential conditions. The absorbing power is determined not merely by the ratio of the air space to the solid material, but by the size of the pores and by the elasticity and viscosity of the mass as a whole. In Fig. 16 Curve 1 is a hair felt, the one alluded to above as of exceptional efficiency. The fraction of its total volume, which is solid material, is 0.12. Curve 2 is a mixture of hair felt and asbestos, whose solid portion is

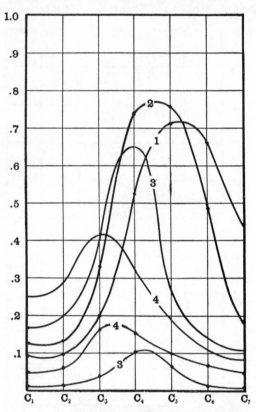

FIG. 18. Curves showing the effect on absorbing power of membrane covering. Curve 1, felt; Curve 2, burlap cemented with silicate of soda; Curve 3, light membrane as described; Curve 4, heavy membrane as described; lower Curve 3, light membrane alone; lower Curve 4, heavy membrane alone.

0.19 of its total volume. Curve 3 is a felt wholly of asbestos $\frac{3}{8}''$ thickness, whose solid portion is 0.33 of its total volume. In this latter the asbestos fiber is felted to an asbestos cloth which serves to strengthen it greatly. Curve 4 is for an asbestos felt without reënforcement. That a considerable fraction of its absorbing power is due to its elastic yielding as a whole is shown by its rather sharp maxima.

The curves in Fig. 17 show the effect of holding the felt at different distances from the wall. In each case it was held on a wire grating. Curve 1 is when the felt is as near the wall as the grating would permit, perhaps within a quarter of an inch of the wall. Curve 2 is when the felt was held at a distance of two inches; Curve 3 at four inches; and Curve 4 at six inches from the wall. It is evident that there is a slight gain from an air space behind the felt, but it is also evident that this gain is so slight as to be entirely incommensurate with the cost of construction and its loss in durability.

The Curves in Fig. 18 show the efficiency of various coverings. Curve 1 is the normal exposed efficiency of the felt above referred to. Curve 2 is its efficiency when covered by burlap attached by silicate of soda. This covering was so sized as to be practically impervious, but was in contact with and a part of the felt. Curves 3 and 4 show the efficiency of coverings which are not in contact with the felt, but which are stretched. Both coverings are impervious, — 3 relatively light, 4 heavy. Number 3 weighs 0.87 ounces to the square foot; number 4 weighs 2.58 ounces to the square foot. The materials of which these coverings are made have no bearing on the question, and would be misleading if stated. The really significant factors are their weight, the tension with which they are stretched, their elasticity, and their viscosity. The weight of the several coverings has been stated; the other factors can be defined best by means of their independent absorbing powers. Lower curves 3 and 4 indicate the absorbing power of the membrane coverings alone. It is interesting to note that the diaphragm which has by itself the least absorbing power has the greatest absorbing power when combined with the felt. This is by no means a paradox. It is exactly the result which could be predicted by application of the simplest of physical principles.

Fig. 1. Roman Theatre at Orange, France.

7

THEATRE ACOUSTICS[1]

VITRUVIUS, De Architectura, Liber V, Cap. VIII. (*De locis consonantibus ad theatra eligendis.*)

"All this being arranged, we must see with even greater care that a position has been taken where the voice falls softly and is not so reflected as to produce a confused effect on the ear. There are some positions offering natural obstructions to the projection of the voice, as for instance the dissonant, which in Greek are termed κατηχοῦντες; the circumsonant, which with them are named περιηχοῦντες; and again the resonant, which are termed ἀντηχοῦντες. The consonant positions are called by them συνηχοῦντες.

The dissonant are those places in which the sound first uttered is carried up, strikes against solid bodies above, and, reflected, checks as it falls the rise of the succeeding sound.

The circumsonant are those in which the voice spreading in all directions is reflected into the middle, where it dissolves, confusing the case endings, and dies away in sounds of indistinct meaning.

The resonant are those in which the voice comes in contact with some solid substance and is reflected, producing an echo and making the case terminations double.

The consonant are those in which the voice is supported and strengthened, and reaches the ear in words which are clear and distinct."

This is an admirable analysis of the problem of theatre acoustics. But to adapt it to modern nomenclature, we must substitute for the word dissonance, interference; for the word circumsonance, reverberation; for the word resonance, echo. For consonance, we have unfortunately no single term, but the conception is one which is fundamental.

It is possible that in the above translation and in the following interpretation I have read into the text of Vitruvius a definiteness of conception and an accord with modern science which his language only fortuitously permits. If so, it is erring on the better side, and is but a reasonable latitude to take under the circumstances. The only passage whose interpretation is open to serious question is that re-

---

[1] The American Architect, vol. civ, p. 257.

lating to dissonant places. If Vitruvius knew that the superposition of two sounds could produce silence, and the expression *"opprimit insequentis vocis elationem"* permits of such interpretation, it must stand as an observation isolated by many centuries from the modern knowledge of the now familiar phenomenon of interference.

INTERFERENCE

Interference is a phenomenon common to all types of wave motion. The best introduction to its discussion is by reference to water-waves

FIG. 2. Greek Theatre at the University of California. Mr. John Galen Howard, Architect.

and in particular to an interesting example of tidal interference on the Tongking Peninsula. The tide of the Pacific Ocean enters the Chinese Sea through two channels, one to the north of the Philippine Islands, between Luzon and Formosa, and the other through the Sulu Archipelago between Mindanao and Borneo. The northern channel is short and deep; and the tide enters with very little retardation. The other channel, although broad, is shallow, tortuous, and broken by many small islands; and the tide in passing through is much retarded. The two tides thus entering the Chinese Sea produce an effect which varies from point to point. At one port on the Tongking Peninsula, these tides are so retarded relatively to each other as to be six hours apart. It is high tide by one when it is low tide by the other. It also so happens that at this point the two tides are equal. Being equal and exactly opposite in phase, they neutralize each other.

Because tidal waves are long in comparison with the bodies of water in which they are propagated, their interference phenomena are obscure except to careful analysis. When, however, the waves are smaller than the space in which they are being propagated, the interference system becomes more marked, more complicated, and more interesting. Under such circumstances, there may be regions of perfect quiet near regions of violent disturbance.

Subjecting the parallel to a more exact statement, whenever two water-waves come together the resulting disturbance at any instant is equal to the algebraic sum of the disturbances which each would produce separately. If their crests coincide, the joint effect is equal to the sum of their separate effect. If crest and trough coincide, their joint effect is the difference between them. If their relative retardation is intermediate, a wave results which is intermediate between their sum and their difference and whose time of maximum does not occur simultaneously with the maximum of either of the components.

The phenomenon is one which may be produced accurately on any scale and with any type of wave motion. Thus sound consists of waves of alternate condensation and rarefaction in the air. If two trains of sound-waves cross each other so that at a given point condensation in the two trains arrive simultaneously, the rarefactions will also arrive simultaneously, and the total disturbance is a train of waves of condensation and rarefaction equal to the sum of the two components. If one train is retarded so that its condensations coincide with the other's rarefactions, the disturbance produced is the difference between that which would be produced by the trains of waves separately. Just as a tidal wave, a storm wave, or a ripple may be made to separate and recross by some obstacle round which it diffracts or from which it is reflected, and recombining produce regions of violent and regions of minimum disturbances, so sound-waves may be diffracted or reflected, and recombining after travelling different paths, produce regions of great loudness and regions of almost complete silence. In general, in an auditorium the phenomenon of interference is produced not by the crossing of two trains of waves only, but by the crossing of many, reflected from the various walls, from the ceiling, from the floor, from any obstacle whatever in the room, while still other trains of waves are produced by the diffraction of the sound around columns and pilasters.

A source of sound on whose steadiness one can rely is all that is necessary in order to make the phenomenon of interference obvious. A low note on a pure toned stop of a church organ will serve the purpose admirably. The observer can satisfy himself that the note is sounding steadily by remaining in a fixed position. As soon, however, as he begins to move from this position by walking up and down the aisle he will observe a great change in loudness. Indeed, he may find a position for one ear which, if he closes the other, will give almost absolute silence, and this not far from positions where the sound is loud to the extent of being disagreeable. The observer in walking about the church will find that the phenomenon is complicated. It is, however, by no means random in its character, but definite, permanent, and accurate in its recurrence, note for note. The phenomenon, while difficult, is by no means impossible of experimental investigation or of theoretical solution. Indeed, this has been done with great care in connection with the study of another problem,—that of the Central Criminal Court Room in London known as Old Bailey. The full primary explanation of the methods and results of this general investigation would be inappropriately long in an article dealing with the acoustics of theatres; for while interference is a factor in every auditorium, it is on the whole not the most seriously disturbing factor in theatre design.

The subject of interference would not have been given even so extended a discussion as this in a paper dealing with theatres were it not that recently there has been proposed in Germany a form of stage setting known as the Kuppel-horizont for sky and horizon effects, to accompany the Fortuny system of stage lighting, in which interference may be a not inconsiderable factor unless guarded against. The Fortuny system, which in the opinion of some competent judges is an effective form of stage lighting, consists primarily in the use of indirect illumination, softened and colored by reflection from screens of silk. As an adjunct to the system, and in an endeavor to secure a considerable depth to the stage without either great height or an excessive use of sky and wing flies, a cupola is recommended to go with the Fortuny lighting as shown in the accompanying figures taken from the publications of the *Berliner Allegemeine Electricitäts Gesellschaft*. In Figs. 3 and 4, the cupola is shown in section and in plan. Lights A and B illuminate the interior of the cupola; C and E light the area of the stage on which the prin-

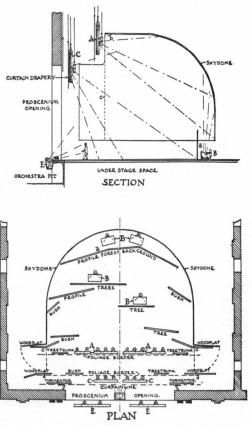

Figs. 3 and 4. Section and plan of the Kuppel-Horizont
with Fortuny system of lighting.

cipal action occurs. Cloud effects, either stationary or moving, are projected on the surface of the cupola by a stereopticon. The great advantage claimed for this form of stage setting is the more natural arrangement of stage properties which it makes possible, and the elimination of numerous flies. On the other hand there is some criticism that this lighting results in an unnatural silhouetting.

So detailed an explanation of the diagrams and the purpose of the several parts is necessitated by the fact that it is as yet an unfamiliar device in this country. It has been introduced recently in a number of theatres in Germany, although I believe not elsewhere, unless possibly in one theatre in England. It has been called to my attention by Professor Baker as a possible equipment of the theatre which

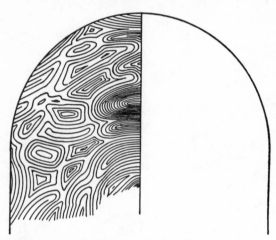

FIG. 5. Interference system for tenor C in the Kuppel-Horizont, having a thirty-six foot proscenium opening. The intensity of sound is represented by contour lines, the maximum variation being forty-seven fold.

is proposed for the dramatic department of Harvard University, and it is reasonable to regard it as a probable factor in theatre design in other countries than Germany.

In Fig. 5 is plotted the interference system established in this space, on a standing head level of five feet from the floor of the stage, by a sustained note tenor C in pitch. The intensity of the sound is indicated by contour lines very much as land elevation is indicated on the maps of the Geodetic Survey. In this plot, account has been taken of the sound reflected from the cupola and from the floor. No account has been taken of the reflection from the walls of the main auditorium since this would be a factor only for sounds prolonged beyond the length of any single element in articulate speech. Even in the case of a very prolonged sound the modification of the interference system of the stage and cupola by the rest of the auditorium would be very slight.

The interference system on the stage in question being determined wholly by the floor and cupola, it may be computed, and in the preparation of the chart was computed, by the so-called method of images. The sound reflected from the floor comes as from a virtual image as far beneath the floor as the mouth of the speaker is above it. Each of these produce real images by reflection from the interior of the cupola. Bearing in mind that these real images show the phenomenon of diffraction and some astigmatism, and taking into account the phase of the sound as determined by reflection and by distance, the calculation is laborious but not difficult. It involves but the most familiar processes of geometrical optics.

The disturbing effect of this interference system is not so great when the speaker is well in front of the center of curvature of the cupola, and of course it is almost always more or less broken by the stage properties, as indicated in Figs. 3 and 4. Nevertheless, it is well to bear in mind that the quarter sphere form, as indicated in the diagrams, is neither necessary from the standpoint of illumination nor desirable from the standpoint of acoustics. Acoustically a flatter back with sharper curvature above and at the sides is preferable.

It should be repeated that the interference system is established only when the tones are sustained, in this case over one-tenth of a second, and is more of an annoyance to the actor on the stage than to the audience. With shorter tones it becomes an echo, and in this form is quite as annoying to the audience as to the actor. It should be added that the interference changes with change of pitch, but preserves extreme maxima and minima for a central position in a spherical or partly spherical surface. Finally in music, since sustained tones occur more than in speech, the interference is more disturbing. The effect of such spherical stage recesses on music is shown by those otherwise unusually excellent auditoriums, Orchestra Hall in Chicago, and the Concert Hall at Willow Grove Park near Philadelphia.

Reverberation

"Circumsonant places" were rare and almost wholly negligible difficulties in Greek and Roman theatres. However, they were common in the temples, and were even more pronounced in some of the older Roman palaces. It must have been in the experience of such conditions, wholly foreign to the theatre of which he was writing, that Vitruvius made this portion of his analysis of the acoustical problem. Given the fundamental form of the Greek theatre, it required no special consideration and little or no skill to avoid such difficulties. However, this is not true of the modern theatre, in which excessive reverberation is more often the defect than any other factor.

If a sound be produced briefly in a wholly empty, wholly closed room, having perfectly rigid walls, it will be reflected at each incidence with undiminished intensity, and, travelling to and fro across the room, will continue audible almost indefinitely. Of course no theatre, ancient or modern, satisfies these conditions and the sound loses at each reflection, diminishing in intensity, until in the course of time it crosses what the experimental psychologist calls the "threshold of audibility." In the Greek theatres the duration of audibility of the residual sound after the cessation of a source of ordinary loudness was never more than a few tenths of a second; in a modern theatre it may be several seconds. The rapidity with which the sound dies away depends on the size of the theatre, on its shape, on the materials used for its walls, ceiling, and furnishings, and on the size and distribution of the audience. The size and shape of the theatre determines the distance travelled by the sound between reflections, while the materials determine the loss at each reflection. No actual wall can be perfectly rigid. Wood sheathing, plaster on wood lath, plaster on wire lath, plaster applied directly to the solid wall, yield under the vibrating pressure of sound and dissipate its energy. Even a wall of solid marble yields slightly, transmitting the energy to external space or absorbing it by its own internal viscosity.

Absorptions by the walls and other objects in the process of reflection, including in this transmission through all openings into outer space as equivalent to total absorption — boundary conditions in other words — are practically alone to be credited with the dissolution of the residual sound. But Vitruvius' statement that the sound "is reflected into the middle, where it dissolves" challenges completeness and at least the mention of another factor, which, because of its almost infinitesimal importance, would otherwise be passed without comment.

Assuming, what is of course impossible, a closed room of absolutely rigid and perfectly reflecting walls, a sound once started would not continue forever, for where the air is condensed by the passing of the wave of sound, it is heated, and where it is rarefied, it is cooled. Between these unequally heated regions and between them and the walls, there is a continual radiation of heat, with a resulting dissipation of available energy. In the course of time, but only in the course of a very long time, the sound would even thus cease to be of audible intensity. This form of dissipation might well be called in the language of Vitruvius "*solvens in medio*"; but, instead of being an important factor, it is an entirely negligible factor in any actual auditorium.

Practically the rapidity with which the sound is absorbed is dependent solely on the nature of the reflecting surfaces and the length of the path which the sound must traverse between reflections, the latter depending on the shape and size of the auditorium. It was shown in a series of papers published in The American Architect in 1900,[1] and in another paper published in the Proceedings of the American Academy of Arts and Sciences in 1906,[2] that, given the plans of an auditorium and the material of which it is composed, it is possible to calculate with a very high degree of accuracy the rate of decay of a sound in the room and the duration of its audibility. In the first of the above papers there was given the complete theory of the subject, together with tables of experimentally determined coefficients of absorption of sound for practically all the materials that enter into auditorium construction, for sounds having a pitch one octave above middle C (vibration frequency 512). In the second of the above papers there were given the coefficients of absorption of building materials for the whole range of the musical scale.

[1] See page 69.   [2] Ibid.

In the careful design of a room for musical purposes, the problem obviously must include the whole range of the musical scale, at least seven octaves. It is not so obvious that the study must cover so great a range when the primary use is to be with the spoken voice. The nearest study to architectural acoustics is the highly developed science of telephony, and in this it is apparently sufficient for much of the work to adapt the theory and design to the single frequency of 800, approximately A in the second octave above middle C. But for

FIG. 6. The Little Theatre, New York. Ingalls and Hoffman, Architects.

some problems the investigation must be extended over a considerable range of pitch. Similarly experience in the architectural problem shows that with some of the materials entering into building construction there occurs a sharp resonance within a not great range of pitch. It is, therefore, necessary to determine the reverberation even for the speaking voice, not for a single pitch but for a considerable range, and the quality of a theatre with respect to reverberation will be represented by a curve in which the reverberation is plotted against the pitch.

Without undertaking to give again a complete discussion of the theory of reverberation, and referring the reader to the earlier (1900) numbers of The American Architect, it will suffice to give a single

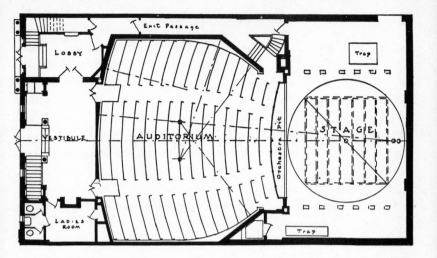

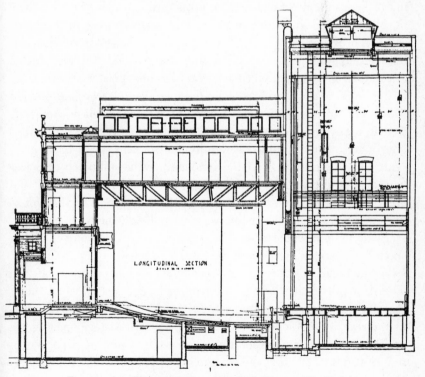

Figs. 7 and 8. Plan and Section of the Little Theatre, New York.
Ingalls and Hoffman, Architects.

illustration. For this I have selected Mr. Winthrop Ames' "Little Theatre" in New York, designed by Messrs. Ingalls and Hoffman, because the purpose and use of this auditorium was defined from the beginning with unusual precision. The purpose was the production of plays which could be adequately rendered only by the most delicate shades of expression, which would be lost in considerable measure if the conditions were such as to necessitate exaggeration of feature or of voice. The definition of its use was that it should seat just less than 300, and that all the seats were to be as nearly as possible of equal excellence, with the important assurance that every seat would be occupied at every performance.

The final plan and section of the Little Theatre are shown in Figs. 7 and 8. The initial pencil sketch was of an auditorium differing in many architectural details, acoustical considerations sharing in, but by no means alone dictating, the steps leading to the final solution of the problem. The first calculations, based on the general lines of the initial sketch, and assuming probable materials and plausible details of construction (plaster on tile walls, plaster on wire lath ceiling, solid plaster cornices and moulding), gave a reverberation as shown in Curve 1 in Fig. 9. This would not have been in excess of that in many theatres whose acoustical qualities are not especially questioned. But the unusual requirements of the plays to be presented in this theatre, and the tendency of the public to criticize whatever is unconventional in design, led both Mr. Ames and the architects to insist on exceptional quality. The floor was, therefore, lowered at the front, the ceiling was lowered, and the walls near the stage brought in and reduced in curvature, with, of course, corresponding changes in the architectural treatment. The rear wall, following the line of the rear seats, remained unchanged in curvature. The side walls near the stage were curved. The net effect of these changes was to give an auditorium 28 feet high in front, 23 feet high at the rear, 48 feet long and 49 feet broad, with a stage opening 18 by 31, and having a reverberation as shown by Curve 2. In order to reduce still further the reverberation, as well as to break acoustically the curvature of the side and rear walls, "acoustic felt" was applied in panels. There were three panels, 6 feet by 13 feet, on each of the side walls, and seven panels, two 4 feet 5 inches by 13 feet, two 5 feet by 10 feet, two 2 feet by 4 feet, and one 8 feet by 7 feet, on the rear wall. The resulting reverberation is shown by Curve 3 in the diagram. Throughout, consideration was had for the actual path of the sound in its successive reflections, but the discussion of this

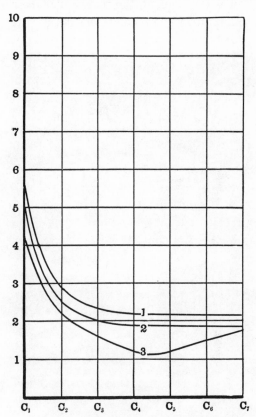

FIG. 9. Reverberation in seconds of the Little Theatre, for notes of different pitch, C_3 being Middle C, Curve 1 for the first design, Curve 2 for the second, and Curve 3 for the third and as built.

phase of the general problem comes in the next section and will be illustrated by other theatres.

It should be said, parenthetically but none the less emphatically, that throughout this paper by theatre is meant an auditorium for the spoken drama.

Echo

When a source of sound is maintained constant for a sufficiently long time — a few seconds will ordinarily suffice — the sound becomes steady at every point in the room. The distribution of the intensity of sound under these conditions is called the interference

Fig. 10.   Interior, the New Theatre, New York.   Carrère and Hastings, Architects.

system, for that particular note, of the room or space in question. If the source of sound is suddenly stopped, it requires some time for the sound in the room to be absorbed. This prolongation of sound after the source has ceased is called reverberation. If the source of sound, instead of being maintained, is short and sharp, it travels as a discrete wave or group of waves about the room, reflected from wall to wall, producing echoes. In the Greek theatre there was ordinarily but one echo, "doubling the case ending," while in the modern theatre there are many, generally arriving at a less interval of time after the direct sound and therefore less distinguishable, but stronger and therefore more disturbing.

This phase of the acoustical problem will be illustrated by two examples, the New Theatre, the most important structure of the

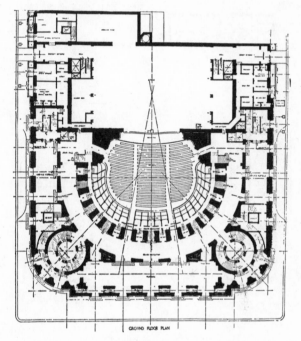

GROUND FLOOR PLAN

Fig. 11

kind in New York, and the plans of the theatre now building for the Scollay Square Realty Company in Boston.

Notwithstanding the fact that there was at one time criticism of the acoustical quality of the New Theatre, the memory of which still lingers and still colors the casual comment, it was not worse in proportion to its size than several other theatres in the city. It is, therefore, not taken as an example because it showed acoustical defects in remarkable degree, but rather because there is much that can be learned from the conditions under which it was built, because such defects as existed have been corrected in large measure, and above all in the hope of aiding in some small way in the restoration of a magnificent building to a dignified use for which it is in so many ways eminently suited. The generous purpose of its Founders, the high ideals of its manager in regard to the plays to be produced, and the perfection otherwise of the building directed an exaggerated and morbid attention to this feature. Aside from the close scrutiny which

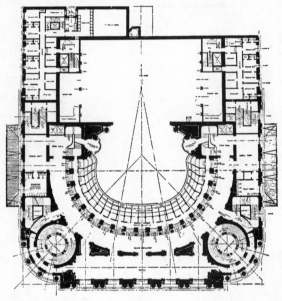

Fig. 12

always centers on a semi-public undertaking, the architects, Messrs. Carrère and Hastings, suffered from that which probably every architect can appreciate from some similar experience of his own, — an impossible program. They were called on to make a large "little theatre," as a particular type of institution is called in England; and, through a division of purpose on the part of the Founders and Advisers, for the Director of the Metropolitan Opera was a powerful factor, they were called on to make a building adapted to both the opera and the drama. There were also financial difficulties, although very different from those usually encountered, a plethora of riches. This necessitated the provision of two rows of boxes, forty-eight originally, equally commodious, and none so near the stage as to thereby suffer in comparison with the others. Finally, there was a change of program when the building was almost complete. The upper row of boxes was abandoned and the shallow balcony thus created was devoted to foyer chairs which were reserved for the

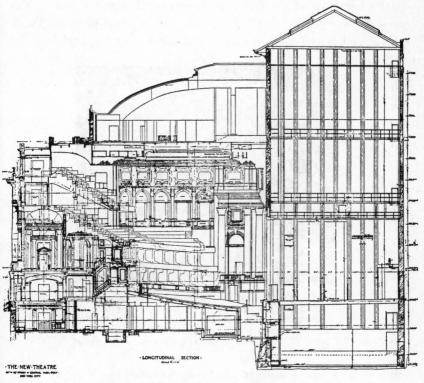

·THE·NEW·THEATRE·

·LONGITUDINAL SECTION·

Fig. 13. Plans and Section of the New Theatre, New York.
Carrère and Hastings, Architects.

annual subscribers. As will be shown later these seats were acoustically the poorest in the house.

Encircling boxes are a familiar arrangement, but most of the precedents, especially those in good repute, are opera houses and not theatres, the opera and the drama being different in their acoustical requirements. In the New Theatre this arrangement exerted a three-fold pressure on the design. It raised the balcony and gallery 12 feet. It increased both the breadth and the depth of the house. And, together with the requirement that these boxes should not extend near the stage, it led to side walls whose most natural architectural treatment was such as to create sources of not inconsiderable echo.

The immediate problem is the discussion of the reflections from the ceiling, from the side walls near the stage, from the screen and parapet in front of the first row of boxes and from the wall at the rear of these boxes. To illustrate this I have taken photographs of the actual sound and its echoes passing through a model of the

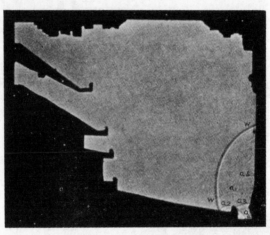

FIG. 14. Photograph of a sound-wave, WW, entering a model of the New Theatre, and of the echoes a_1, produced by the orchestra screen, a_2 from the main floor, a_3, from the floor of the orchestra pit, a_4, the reflection from the orchestra screen of the wave a_3, a_5 the wave originating at the edge of the stage.

theatre by a modification of what may be called the Toeppler-Boys-Foley method of photographing air disturbances. The details of the adaptation of the method to the present investigation will be explained in another paper. It is sufficient here to say that the method consists essentially of taking off the sides of the model, and, as the sound is passing through it, illuminating it instantaneously by the light from a very fine and somewhat distant electric spark. After passing through the model the light falls on a photographic plate placed at a little distance on the other side. The light is refracted by the sound-waves, which thus act practically as their own lens in producing the photograph.

In the accompanying illustrations reduced from the photographs the enframing silhouettes are shadows cast by the model, and all

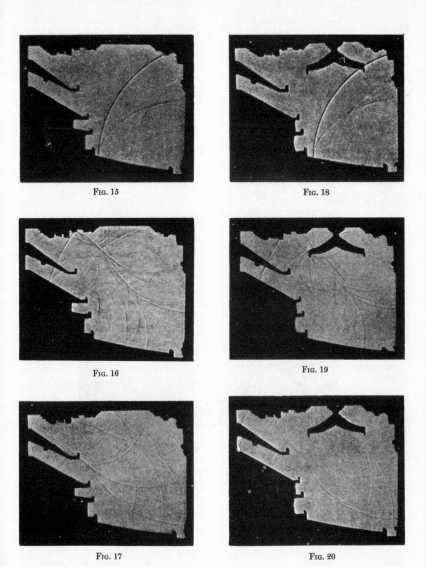

Fig. 15

Fig. 18

Fig. 16

Fig. 19

Fig. 17

Fig. 20

Two series of photographs of the sound and its reflections in the New Theatre, — 15 to 17 before, 18 to 20 after the installation of the canopy in the ceiling. The effect of the canopy in protecting the balcony, foyer chairs, boxes, and the orchestra chairs back of row L is shown by comparing Figs. 19 and 20 with Figs. 16 and 17.

within are direct photographs of the actual sound-wave and its echoes. For example, Fig. 14 shows in silhouette the principal longitudinal section of the main auditorium of the New Theatre. *WW* is a photograph of a sound-wave which has entered the main auditorium from a point on the stage at an ordinary distance back of the proscenium arch; a_1, is the reflection from the solid rail in front of the orchestra pit, and a_2, the reflection from the floor of the sound which has passed over the top of the rail; a_3 is the reflection from the floor

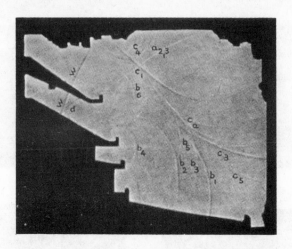

FIG. 21. Photograph of the direct sound, *WW*, and of the echoes from the various surfaces; $a_{2,3}$, a wave, or echo, due to the combination of two waves which originated at the orchestra pit; c_1 from the oval panel in the ceiling; c_a and c_3, from the ceiling mouldings and cornice over the proscenium arch; c_4, a group from the moulding surrounding the panel; c_5, from the proscenium arch; b_1, b_2, b_6 from the screens in front, and the walls in the rear of the boxes, balcony and gallery.

of the pit, and a_4 the reflection of this reflected wave from the rail; while a_5 originated at the edge of the stage. None of these reflections are important factors in determining the acoustical quality of the theatre, but the photograph affords excellent opportunity for showing the manner in which reflections are formed, and to introduce the series of more significant photographs on page 181.

Figures 15, 16, and 17 show the advance of the sound through the auditorium at .07, .10, and .14 second intervals after its departure from the source. In Fig. 15, the waves which originated at the orchestra pit can be readily distinguished, as well as the nascent waves where the primary sound is striking the ceiling cornice immediately over the proscenium arch. The proscenium arch itself was very well designed, for the sound passed parallel to its surface. Otherwise reflections from the proscenium arch would also have shown in the photograph. These would have been directed toward the audience and might have been very perceptible factors in determining the ultimate acoustical quality.

The system of reflected waves in the succeeding photograph in the series is so complicated that it is difficult to identify the several reflections by verbal description. The photograph is, therefore, reproduced in Fig. 21, lettered and with accompanying legends. It is interesting to observe that all the reflected waves which originated at the orchestra pit have disappeared with the exception of waves a_2 and a_3. These have combined to form practically a single wave. Even this combined wave is almost negligible.

The acoustically important reflections in the vertical section are the waves c_1, c_2, and c_3. The waves b_1 and b_2 from the screen in front of the boxes and from the back of the boxes are also of great importance, but the peculiarities of these waves are better shown by photographs taken vertically through a horizontal section.

The waves c_1, c_2, c_3, and b_1 and b_2 show in a striking manner the fallacy of the not uncommon representation of the propagation of sound by straight lines. For example, the wave c_1 is a reflection from the oval panel in the ceiling. The curvature of this panel is such that the ray construction would give practically parallel rays after reflection. Were the geometrical representation by rays an adequate one the reflected wave would thus be a flat disc equal in area to the oblique projection of the panel. As a matter of fact, however, the wave spreads far into the geometrical shadow, as is shown by the curved portion reaching well out toward the proscenium arch. Again, waves c_a and c_3 are reflections from a cornice whose irregularities are not so oriented as to suggest by the simple geometrical representation of rays the formation of such waves as are here clearly shown. But each small cornice moulding originates an almost hemispherical wave, and the mouldings are in two groups, the position of each being such that the spherical waves conspire to form these two master waves. The inadequacy of the discussion of the subject of architectural acoustics by the construction of straight lines is still further shown by the waves reflected from the screens in front of the boxes, of the balcony, and of the gallery. These reflecting surfaces are narrow, but give, as is clearly seen in the photograph, highly divergent waves. This spreading of the wave beyond the geometrical projection is more pronounced the smaller the opening or the reflecting obstacle and the greater the length of the wave. The phenomenon is called diffraction and is, of course, one of the well-known phenomena of physics. It is more pronounced in the long waves of sound than in the short waves of light, and on the small areas of an auditorium than in the large dimensions of out-of-door space. It cannot be ignored, as it has been heretofore ignored in all discussion of this phase of the problem of architectural acoustics, with impunity. The method of rays, although a fairly correct approximation with large areas, is misleading under most conditions. For example, in the present case it would have predicted almost perfect acoustics in the boxes and on the main floor.

Figures 17 and 20 show the condition in the room when the main sound-wave has reached the last seat in the top gallery. The wave c_1 has advanced and is reaching the front row of seats in the gallery, producing the effect of an echo. A little later it will enter the balcony, producing there an echo greater in intensity, more delayed, and affecting more than half the seats in the balcony, for it will curve under the gallery, in the manner just explained, and disturb seats which geometrically would be protected. Still later it will enter the foyer seats and the boxes. But the main disturbance in these seats and the boxes, as is well shown by the photograph, arises from the wave c_2, and in the orchestra seats on the floor from the wave c_3.

In the summer following the opening of the theatre, a canopy, oval in plan and slightly larger than the ceiling oval, was hung from the ceiling surrounding a central chandelier. The effect of this in preventing these disturbing reflections is shown by a comparison, pair by pair, of the two series of photographs, Figs. 15 to 17 and Figs. 18 to 20. It is safe to say that there are few, possibly no modern theatres, or opera houses, equal in size and seating capacity,

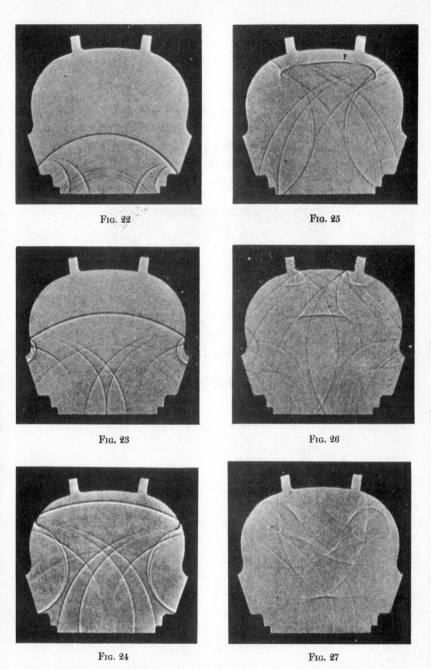

Fig. 22

Fig. 25

Fig. 23

Fig. 26

Fig. 24

Fig. 27

Photographs showing the reflections, in a vertical plane, from the sides of the proscenium arch, the plain wall below the actors' box, and the rail or screen in front of the boxes. The photographs taken in numerical sequence show the progress of a single sound-wave and its reflections.

which are so free from this particular type of disturbance as the New Theatre at the present time.

In the study of the New Theatre, photographs were taken through several horizontal sections. It will be sufficient for the purposes of the present paper to illustrate the effect of curved surfaces in producing converging waves by a few photographs showing the propagation of sound through a single section in a plane passing through the parapet in front of the boxes. The reflected waves shown in

FIG. 28. A photograph, one of many taken, showing in vertical section one stage of the reflection b_2, Fig. 21. These reflections were eliminated by the architects in the summer following the opening of the theatre, but have been in part restored by subsequent changes.

Fig. 22 originating from the edge of the proscenium arch and from the base of the column can be followed throughout all the succeeding photographs. In Fig. 23 are shown waves originating from the plain wall beneath the actor's box and the beginning of some small waves from the curved parapet. It is easily possible, as it is also interesting and instructive, to follow these waves through the succeeding photographs. In Fig. 25 the sound has been reflected from the rear of the parapet; while in Fig. 26 it has advanced further down the main floor of the auditorium, narrowing as it proceeds and gaining in intensity. The waves reflected from the parapet outside of the aisles are here shown approaching each other behind the wave which has been reflected from the parapet between the aisles. Waves are also shown in Fig. 26 emerging from the passages between the boxes.

Indeed, it is possible to trace the waves arising from a second reflection from the proscenium arch of the sound which, first reflected from the corresponding surfaces on the other side, has crossed directly in front of the stage. With a little care, it is possible also to identify these waves in the last photograph.

Although many were taken, it will suffice to show a single photograph, Fig. 28, of the reflections in the plane passing through the back of the boxes. These disturbing reflections were almost entirely eliminated in the revision of the theatre by the removal of the boxes from the first to the second row and by utilizing the space vacated together with the anterooms as a single balcony filled with seats.

An excellent illustration of the use of such photographs in planning, before construction and while all the forms are still fluid, is to be found in one of the theatres now being built in Boston by Mr. C. H. Blackall, who has had an exceptionally large and successful experience in theatre design. The initial pencil sketch, Fig. 29, gave in the model test the waves shown in the progressive series of photographs, Figs. 31 to 33. The ceiling of interpenetrating cylinders was then changed to the form shown in finished section in Fig. 30, with the results strikingly indicated in the parallel series of photographs, Figs. 34 to 36. It is, of course, easy to identify all the reflections in each of these photographs,—the reflections from the ceiling and the balcony front in the first; from the ceiling and from both the balcony and gallery front in the second; and in the third photograph of the series, the reflections of the ceiling reflection from the balcony and gallery fronts and from the floor. But the essential point to be observed, in comparing the two series pair by pair, is the almost total absence in the second series of the ceiling echo and the relatively clear condition back of the advancing sound-wave.

CONSONANCE

Consonance is the process whereby, due to suitably placed reflecting walls, "the voice is supported and strengthened." It is the one acoustical virtue that is positive. It is also the characteristic virtue of the modern theatre, and that through which this complicated auditorium surmounts the attendant evils of interference, reverberation, and echo. Yet such is our modern analysis of the problem that we do not even have for it a name. On the other hand, it is the virtue which the Greek theatre has in least degree. It is, therefore, all the more interesting that it should have been included in the analysis of Vitruvius, and should have received a name so accurately descriptive. Indeed, one can hardly make explanation of the phenomenon better than through the very type of theatre in which its lack is the one admitted defect.

The Greek theatre enjoys a not wholly well-founded reputation for extremely good acoustics. In most respects it is deserved; but

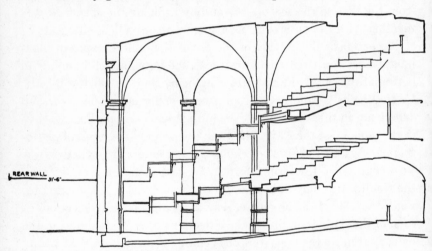

REAR WALL 31'-6"

Fig. 29. Section in pencil sketch of Scollay Square Theatre, Boston.
Mr. C. H. Blackall, Architect.

the careful classical scholar, however gratified he may be by this praise of a notable Greek invention, regards himself as barred by contemporaneous evidence from accepting for the theatre unqualified praise. Every traveler has heard of the remarkable quality of these theatres, and makes a trial wherever opportunity permits, be it at beautiful Taormina, in the steep sloped theatre at Pompeii, the great theatre at Ephesus, or the "little theatre" on the top of Tusculum,— always with gratifying results and the satisfaction of having confirmed a well-known fact. Perhaps it is useless to try to traverse such a test. But there is not a theatre in Italy or Greece which is not in so ruined a condition today that it in no way whatever resembles acoustically its original form. If its acoustics are perfect today, they certainly were not originally. Complete "scaena" and enclosing walls distinctly altered the acoustical conditions. The traveler has in general tested what is little more than a depression in the ground, or a hollow in a quiet country hillside. As a matter of fact, the theatre in its original form was better than in its ruined state. Still, with all its excellencies it was not wholly good. Its acoustical qualities were not wholly acceptable to its contemporaries,

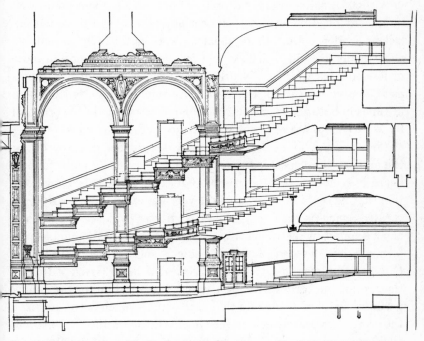

FIG. 30. Finished section of Scollay Square Theatre, Boston. Mr. C. H. Blackall, Architect.

and would be less acceptable in a modern theatre, and for modern drama.

The difficulty with such casual evidence is that it is gathered under wholly abnormal conditions. Not only are the ruins but scant reminders of the original structure, but the absence of a large audience vitiates the test, as it would vitiate a test of any modern theatre. But while in a modern auditorium the presence of an audience almost always, though not invariably, improves the acoustics, in the classical theatre the presence of an audience, in so far as it has any effect, is disadvantageous. The effect of an audience is always twofold, — it diminishes the reverberation, and it diminishes the loudness or intensity of the voice. In general, the one effect is advantageous, the other disadvantageous. But in the Greek theatre, occupied or unoccupied, ruined or in its original form, there was very little reverberation. In fact, this was its merit. On the other hand, the very fact that there was little reverberation is significant that there was very slight architectural reënforcement of the voice. One might well be unconvinced by such *a priori* considerations were there not excellent evidence that these theatres were not wholly acceptable acoustically even in their day, and for drama written for and more or less adapted to them. Excellent evidence that there was insufficient consonance is to be found in the megaphone mouthpieces used at times in both the tragic and the comic masks, and in the proposal by Vitruvius to use resonant vases to strengthen the voice.

The doubt is not as to whether a speaker, turned directly toward the audience and speaking in a sustained voice, could make himself heard in remote parts of a crowded Greek theatre. It is almost certain that he could do so, even in the very large and more nearly level theatres, such as the one at Ephesus. Better evidence of this than can be found in the casual test of a lonely ruin is the annual performance by the staff of the Comédie Française in the theatre at Orange. But even this, the best preserved of either Greek or Roman theatres, is but a ruin, and its temporary adaptation for the annual performance is more modern than classical. A much better test is in the exercises regularly held in the Greek Theatre of the University of California, designed by Mr. John Galen Howard, of which President Wheeler speaks in most approving terms. The drama, especially modern drama, differs from sustained speech and formal address in its range of utterance, in modulation, and above all in the requirement that at times it reaches the audience with great dynamic quality but without strain in enunciation. Mere distinctness is not sufficient. It was through a realization of this that the megaphone mouthpiece was invented,—awkward in use and necessarily destructive of many of the finer shades of enunciation. That it was only occasionally used proves that it was not a wholly satisfactory device, but does not detract its evidence of weakness in the acoustics of the theatre.

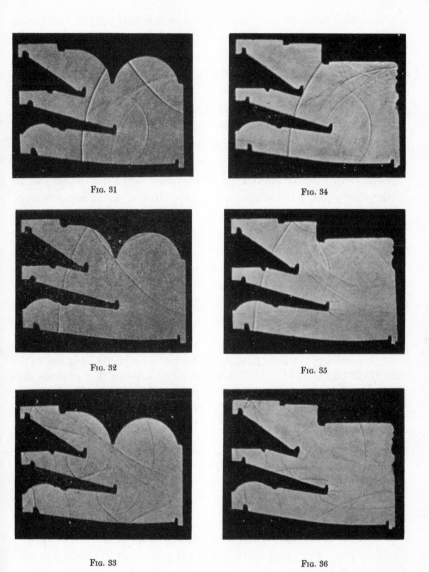

FIG. 31   FIG. 34

FIG. 32   FIG. 35

FIG. 33   FIG. 36

Two series of photographs showing, Figs. 31–33, the reflections which would have resulted from the execution of the first pencil sketch of the Scollay Square Theatre (Fig. 29), and, Figs. 34–36, from the execution of the second sketch by Mr. Blackall (shown in finished section in Fig. 30).

The megaphone mouthpiece bears to the acoustics of the Greek theatre the same evidence, only in a reciprocal form, that the mask itself bears to the theatre's illumination. It was not possible to see in bright daylight, particularly in the bright sunlight of the Mediterranean atmosphere, with anything like the accuracy and detail possible in a darkened theatre with illuminated stage. The pupil of the eye was contracted, and the sensitiveness of the retina exhausted by the brilliancy of the general glare. Add to this that the distance from the stage was very much greater in the Greek than in the modern theatre, audience for audience, and one can realize the reason for the utter impossibility of facial expression in Greek dramatization except by artificial exaggeration. The heaviness and inflexibility of these devices, and, therefore, their significance as proof of some inherent difficulty in dramatic presentation, is emphasized by the delicacy of line and fine appreciation of the human form shown in other contemporaneous art.

Not less significant in regard to the acoustics of the Greek theatre are the directions given by Vitruvius for the reënforcement of the voice by the use of resonant vases:

" Accordingly bronze vessels should be made, proportional in size to the size of the theatre, and so fashioned that when sounded they produce with one another the notes of the fourth, the fifth, and so on to the double octave. These vessels should be placed in accordance with musical laws in niches between the seats of the theatre in such position that they nowhere touch the wall, but have a clear space on all sides and above them. They should be set upside down and supported on the side facing the stage by wedges not less than half a foot high. . . . With this arrangement, the voice, spreading from the stage as a center, and striking against the cavities of the different vessels, will be increased in volume and will wake an harmonious note in unison with itself."

There is good reason for believing that this device was but very rarely tried. This, and the fact that it could not possibly have accomplished the purpose as outlined by Vitruvius, is not germane. The important point is that its mere proposal is evidence that the contemporaries of the Greek theatres were not wholly satisfied, and that the defect was in lack of consonance.

It would be inappropriately elaborate and beyond the possible length of this paper to give in detail the method of calculating the loudness of sound in different parts of an auditorium. That subject is reserved for another paper in preparation, in which will be given not merely the method of calculation but the necessary tables for its simplification. It is, however, possible and proper to give a general statement of the principles and processes involved.

In this discussion I shall leave out as already adequately discussed the phenomenon of interference, or rather shall dismiss the subject with a statement that when two sounds of the same pitch are super-posed in exact agreement of phase, the intensity of the sound is the square of the sum of the square roots of their separate intensities; when they are in opposite phases, it is the square of the difference of the square roots of their intensities; but when several sounds of the same pitch arrive at any point in the room with a random difference of phase their probable intensity is the simple numerical sum of their separate intensities. It is on the assumption of a random difference of phase and an average probable loudness that I shall here consider the question. This has the advantage of being the simpler and also a first approximation in an auditorium designed for articulate speech.

When sound spreads from a spherically symmetrical source it diminishes as the square of the distance. When the sound is being propagated, still in space unrestricted by walls or ceiling, but over the heads of a closely seated audience, the law of the diminution of the sound is more rapid than the law of the inverse square. This more rapid diminution of the sound is due to the absorption of the sound by the audience. It is a function of the elevation of the speaker and the angle of inclination of the floor,— in other words, the angle between the sight lines. The diminution of the intensity of the sound due to distance is less the greater this angle.

If the auditorium be enclosed by not too remote walls, the voice coming directly from the speaker is reënforced by the reflection from the retaining walls. However, it is obvious that the sounds reflected from the walls and ceilings have traversed greater paths than the sound of the voice which has come directly. If this difference of path length is great, the sounds will not arrive simultaneously. If, how-ever, the path differences are not great, the reflected sounds will arrive in time to reënforce the voice which has come directly, each syllable by itself, or, indeed, in time for the self support of the subsyllabic components. It is to this mutual strengthening of concurrent sounds within each element of articulate speech that Vitruvius has given the name "consonance."

Thus in the computation of the intensity of the voice which has come directly from the speaker across the auditorium, it is necessary to take into consideration not merely the diminution of intensity according to the law of the inverse square of the distance and the diminution of the intensity due to the absorption by the clothing of

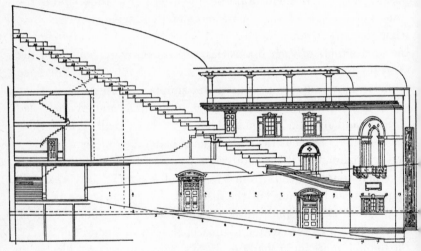

Fig. 37. The Harris Theatre, Minneapolis, first design.
Chapman and Magney, Architects.

the audience, but also, as a compensating factor for the latter, the diffraction of the sound from above which is ever supplying the loss due to absorption, while in computing the intensity of the sound reflected from any wall or other surface one must take into consideration all this, and also the coefficient of reflection of the wall and the diffraction due to the restricted area of the reflecting element.

Abstract principles are sometimes tedious to follow even when not difficult. In Fig. 38 is shown a photograph taken in an investigation for the architects, Messrs. Chapman and Magney, of the Harris Theatre, to be erected in Minneapolis, which affords an excellent example of both favorable and unfavorable conditions in respect to consonance. The initial sketch for this theatre offered no problems either of interference or reverberation, and of echo only in the horizontal section. The only very considerable question presented by the plans was in respect to consonance and there in regard only to the more remote parts of the floor and of the balcony. The particular photograph here reproduced records the condition of the sound in the room at such an instant as to bring out this aspect of the problem in marked degree.

The forward third of the balcony in this theatre affords an excellent example of consonance, for the reflection from the ceiling arrives so nearly simultaneously with the sound which has come

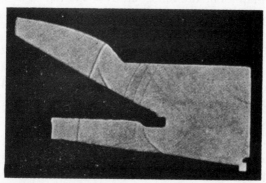

FIG. 38. Showing the consonance in the balcony of the Harris Theatre. This relates only to consonance in the vertical section.

directly from the stage as to "strengthen and support" it and yet "leave the words clear and distinct." The interval between the two, the direct and the reflected voice, varies from .01 second to .03 second. Back of the first third, however, the consonance from the ceiling gradually diminishes and is practically imperceptible beyond the middle of the gallery. Back of that point the direct voice diminishes rapidly since it is passing in a confined space over the highly absorbent clothing of the audience. The loss of intensity at the rear of the gallery is increased by the carrying of the horizontal portion of the ceiling so far rearward. While the effect of this is to throttle the rear of the gallery it obviously strengthens the voice in the forward third. Although there is thus some compensation, on the whole the forward part of the gallery does not need this service so

much as the rear seats. The photograph shows this process clearly:
the main sound-wave can be seen advancing after having passed the
angle in the ceiling. The wave reflected from the ceiling can be seen
just striking the gallery seats. It is evident that at the instant at
which the photograph was taken the sound-wave was receiving the
last of this support by the sound reflected from the ceiling.

The photograph also shows how the sound after passing the ceiling angle spreads into the space above, thus losing for the moment
thirty per cent of its intensity, a loss, however, to be regained in
considerable part later.

On the main floor the reflection from the ceiling strengthens the
direct voice only for the long syllabic components. Nevertheless, in
comparison with other theatres the forward part of the floor of this
theatre will be excellent. There will be just a trace of echo immediately under the front of the balcony, but this will be imperceptible
beyond the first four rows of seats under the balcony. It is obvious
from the photograph that there is no consonance in the rear of the
main floor of the auditorium under the balcony.

A not unnatural, certainly a not uncommon, inquiry is for some
statement of the best height, the best breadth, and the best depth for
a theatre, for a list of commended and a list of prohibited forms and
dimensions. A little consideration, however, will show that this is
neither a possible nor the most desirable result of such an investigation.

For a simple rectangular auditorium of determined horizontal
dimensions there is a best height. When, however, the horizontal
dimensions are changed the desirable height changes, although by no
means proportionally. When the floor is inclined, when the walls are
curved, when there are galleries and connection corridors, when the
material of construction is varied in character, the problem becomes
somewhat more intricate, the value of each element being dependent
on the others. Moreover it is futile to attempt to formulate a standard form even of a single type of auditorium. How greatly the
design must vary is well illustrated in the four theatres which have
been taken as examples, — the Little Theatre with all the seats on
the main floor, the Harris Theatre, very long, very broad, and with

but a single gallery, the Scollay Square Theatre with two galleries, and the New Theatre with two rows of boxes and two galleries. The fundamental conditions of the problem, not the entirely free choice of the architect, determined the general solution in each case. Acoustical quality is never the sole consideration; at best it is but a factor, introduced sometimes early, sometimes late, into the design.

8

BUILDING MATERIAL AND MUSICAL PITCH [1]

THE absorbing power of the various materials that enter into
the construction and furnishing of an auditorium is but one phase
in the general investigation of the subject of architectural acoustics
which the writer has been prosecuting for the past eighteen years.
During the first five years the investigation was devoted almost
exclusively to the determination of the coefficients of absorption
for sounds having the pitch of violin C (512 vibrations per second).
The results were published in the American Architect and the Engineering Record in 1900.[2]   It was obvious from the beginning that
an investigation relating only to a single pitch was but a preliminary
excursion, and that the complete solution of the problem called for
an extension of the investigation to cover the whole range in pitch
of the speaking voice and of the musical scale.   Therefore during
the years which have since elapsed the investigation has been extended over a range in pitch from three octaves below to three
octaves above violin C.   That it has taken so long is due to the fact
that other aspects of the acoustical problem also pressed for solution, such for example as those depending on form, — interference,
resonance, and echo.   The delay has also been due in part to the
nature of the investigation, which has necessarily been opportunist
in character and, given every opportunity, somewhat laborious and
exhausting.   Some measure of the labor involved may be gained
from the fact that the investigation of the absorption coefficients
for the single note of violin C required every other night from twelve
until five for a period of three years.

While many improvements have been made in the methods of
investigation and in the apparatus employed since the first paper
was published fourteen years ago, the present paper is devoted solely
to the presentation of the results.   I shall venture to discuss, although briefly, the circumstances under which the measurements

[1] The Brickbuilder, vol. xxiii, no. 1, January, 1914.   [2] No. 1, p. 1.

were made, my object being to so interest architects that they will call attention to any opportunities which may come to their notice for the further extension of this work; for, while the absorbing powers of many materials have already been determined, it is evident that the list is still incomplete. For example, the coefficient of glass has been determined only for the note first studied, C, an octave above middle C. In 1898 the University had just completed the construction of some greenhouses in the Botanical Gardens, which, before the plants were moved in, fulfilled admirably the conditions necessary for accurate experimenting. Glass formed a very large part of the area of the enclosing surfaces, all, in fact, except the floor, and this was of concrete whose coefficient of absorption was low and had already been determined with accuracy. By this good fortune it was possible to determine the absorbing power of single-thickness glass. But at that time the apparatus was adapted only to the study of one note; and as the greenhouse was soon fully occupied with growing plants which could not be moved without danger, it was no longer available for the purpose when the scope of the investigation was extended. Since then no similar or nearly so good opportunity has presented itself, and the absorbing power of this important structural surface over the range of the musical scale has not as yet been determined. There was what seemed for the moment to be an opportunity for obtaining this data in an indoor tennis court which Messrs. McKim, Mead and White were erecting at Rhinebeck on the Hudson, and the architects undertook to secure the privilege of experimenting in the room, but inquiry showed that the tennis court was of turf, the absorption of which was so large and variable as to prevent an accurate determination of the coefficients for the glass. The necessary conditions for such experiments are that the material to be investigated shall be large in area, and that the other materials shall be small in area, low in power of absorption, and constant in character; while a contributing factor to the ease and accuracy of the investigation is that the room shall be so located as to be very quiet at some period of the day or night. The present paper is, therefore, a report of progress as well as an appeal for further opportunities, and it is hoped that it will not be out of place at the end of the paper to point out some of the problems which remain and ask that interested architects call attention to any rooms in which it may be possible to complete the work.

The investigation does not wholly wait an opportunity. A special room, exceptionally well adapted to the purpose in size, shape, and location, has been constantly available for the research in one form or another. This room, initially lined with brick set in cement, has been lined in turn with tile of various kinds, with plaster, and with plaster on wood lath, as well as finished from time to time in other surfaces. This process, however, is expensive, and carried out in completeness would be beyond what could be borne personally. Moreover, it has further limitations. For example, it is not possible in this room to determine the absorbing power of glass windows, for one of the essential features of a window is that the outside space to which the sound is transmitted shall be open and unobstructed. An inner lining of glass, even though this be placed several inches from the wall, would not with certainty represent normal conditions or show the effect of windows as ordinarily employed in an auditorium. Notwithstanding these limitations, this room, carefully studied in respect to the effects of its peculiarities of form, especially such as arise from interference and resonance, has been of great service.

Wall and Ceiling-Surfaces

It is well to bear in mind that the absorption of sound by a wall-surface is structural and not superficial. That it is superficial is one of the most widespread and persistent fallacies. When this investigation was initially undertaken in an endeavor to correct the acoustics in the lecture-room of the Fogg Art Museum, one of the first suggestions was that the walls were too smooth and should be roughened. The proposal at that time was that the walls be re-plastered and scarred with the toothed trowel in a swirling motion and then painted, a type of decoration common twenty years ago. A few years later inquiries were received in regard to sanded surfaces, and still later in regard to a rough, pebbly surface of un-troweled plaster; while within the past three years there have been many inquiries as to the efficiency of roughened brick or of rough hewn stone. On the general principle of investigating any proposal so long as it contained even a possibility of merit, these suggestions were put to test. The concrete floor of a room was covered with a gravel so sifted that each pebble was about one-eighth of an inch in diameter. This was spread over the floor so that pebble touched pebble, making a layer of but a single pebble in thickness. It showed not the slightest absorbing power, and there was no perceptible decrease in reverberation. The room was again tried with sand. Of course, it was not possible in this case to insure the thickness of a single grain only, but as far as possible this was accomplished. The result was the same. The scarred, the sanded, the pebbly plaster, and the rough hewn stone are only infinitesimally more efficient as absorbents than the same walls smooth or even polished. The failure of such roughening of the wall-surfaces to increase either the absorption or the dispersion of sound reflected from it is due to the fact that the sound-waves, even of the highest notes, are long in comparison with the dimensions of the irregularities thus introduced.

The absorption of sound by a wall is therefore a structural phenomenon. It is almost infinitely varied in the details of its mechanism, but capable of classification in a few simple modes. The fundamental process common to all is an actual yielding of the wall-surface to the vibrating pressure of the sound. How much the wall yields and what becomes of the motion thus taken up, depends on the nature of the structure. The simplest type of wall is obviously illustrated by concrete without steel reënforcement, for in this there is the nearest approach to perfect homogeneity. The amount that this wall would yield would depend upon its dimensions, particularly its thickness, and upon the density, the elasticity, and the viscosity of the material. It is possible to calculate this directly from the elements involved, but the process would be neither interesting nor convincing to an architect. It is in every way more satisfactory to determine the absorbing power by direct experiment. A concrete wall was not available. In its stead, the next more homogeneous wall was investigated, an eighteen-inch wall of brick set in cement. This wall was a very powerful reflector and its absorbing power exceedingly slight. Without going

into the details of the experiment, it will suffice here to say that
this wall absorbed one and one-tenth per cent of the lowest note
investigated, a C two octaves below middle C, having a vibration
frequency of sixty-four per second; one and two-tenths per cent
of sounds an octave in pitch higher; one and four-tenths per cent
of sounds of middle C; one and seven-tenths per cent for violin C;
two per cent for sounds having a pitch one octave above; two and
three-tenths for two octaves above; and two and one-half per cent
for sounds having a pitch three octaves above violin C, that is to
say, 4094 vibrations per second, the highest note investigated.
These may be written as coefficients of absorption thus:

C_1, .011; C_2, .012; C_3, .014; C_4, .017; C_5, .020; C_6, .023; C_7, .025.

There is a graphical method of presenting these results which is
always employed in physics, and frequently in other branches of
science, when the phenomenon under investigation is simply progressive and dependent upon a single variable. Whenever these
conditions are satisfied — and they are usually satisfied in any
well conducted investigation — the graphical representation of
the results takes the form of a diagram in which the results of the
measurements are plotted vertically at horizontal distances determined by the variable condition. Thus in the following diagram
(Curve 1, Fig. 1) the coefficients of absorption are plotted vertically,
the varying pitch being represented by horizontal distances along the
base line. Such a diagrammatic representation serves to reveal the
accuracy of the work. If the phenomenon is a continuous one,
the plotted points should lie on a smooth curve; the nearness with
which they do so is a measure of the accuracy of the work if the
points thus plotted are determined by entirely independent experiments. This form of diagrammatic representation serves another
purpose in permitting of the convenient interpolation for values
intermediate between observed values. The coefficients for each
type of wall-surface will be given both numerically and diagram-
matically. In order to avoid confusion, the observed points have
been indicated only on the curve for wood sheathing in Fig. 1. It
will suffice to say merely that the other curves on this diagram
are drawn accurately through the plotted observations.

The next wall-surface investigated was plaster on hollow terra cotta tile. The plaster coat was of gypsum hard plaster, the rough plaster being five-eighths of an inch in thickness. The result shows a slightly greater absorption due to the greater flexibility of a hollow

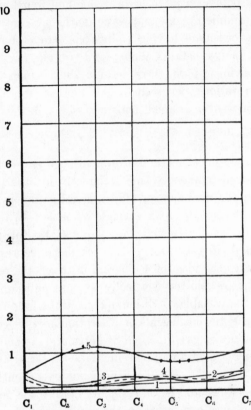

FIG. 1. Absorbing power for sounds varying in pitch from C = 64 to C = 4,096: 1, brick wall; 2, plaster on terra cotta hollow tile; 3, plaster on wire lath; 4, same with skim coat; 5, wood sheathing.

tile wall rather than to any direct effect of the plaster. The difference, however, is not great. The numerical results are as follows (Curve 2, Fig. 1):

C_1, .012; C_2, .013; C_3, .015; C_4, .020; C_5, .028; C_6, .040; C_7, .050.

C_1 is the lowest note, 64 vibrations per second; C_7, the highest, 4,096 per second; the other notes at octave intervals between.

Plaster on an otherwise homogeneous sustaining wall is a first step in the direction of a compound wall, but a vastly greater step is taken when the plaster instead of being applied directly to the sustaining wall is furred to a greater or less distance. In a homogeneous wall, the absorption of sound is partially by communication of the vibration to the material of the wall, whence it is telephoned throughout the structure, and partly by a yielding of the wall as a whole, the sound being then communicated to outside space. In a compound wall in which the exposed surface is furred from the main structure of the wall, the former vibrates between the furring strips like a drum. Such a surface obviously yields more than would a surface of plaster applied directly to tile or brick. The energy which is thus absorbed is partly dissipated by the viscosity of the plaster, partly by transmission in the air space behind it, and partly through the furring strips to the main wall. The mechanism of this process is interesting in that it shows how the free standing plaster may absorb a great amount of sound and may present a greater possibility of resonance and of selective absorption in the different registers of pitch. It is obvious that we are here dealing with a problem of more complicated aspect. It is conceivable that the absorption coefficient should depend on the nature of the supporting construction, whether wood lath, wire lath, or expanded metal lath; on the distance apart of the studding, or the depth of the air space; or, and even more decidedly, on the nature of the plaster employed, whether the old lime plaster or the modern quick setting gypsum plaster. A start has been made on a study of this problem, but it is not as yet so far advanced as to permit of a systematic correlation of the results. It must suffice to present here the values for a single construction. The most interesting case is that in which lime plaster was applied to wood lath, on wood studding at fourteen-inch spacing, forming a two-inch air space. The coefficients of absorption before the finishing coat was put on were (Curve 3, Fig. 1):

C_1, .048; C_2, .020; C_3, .024; C_4, .034; C_5, .030; C_6, .028; C_7, .043.

The values after the finishing coat was put on were as follows (Curve 4, dotted, Fig. 1):

C_1, .036; C_2, .012; C_3, .013; C_4, .018; C_5, .045; C_6, .028; C_7, .055.

It should be remarked that the determination of these coefficients was made within two weeks after the plaster was applied and also that the modern lime is not the same as the lime used thirty years ago, either in the manner in which it is handled or in the manner in which it sets and dries. It is particularly interesting to note in these observations, more clearly in the plotted curves, the phenomenon of resonance as shown by the maxima, and the effect of the increased thickness produced by the skim coat in increasing the rigidity of the wall, decreasing its absorbing power, and shifting the resonance.

The most firmly established traditions of both instrumental and architectural acoustics relate to the use of wood and excite the liveliest interest in the effect of wood sheathing as an interior surface for auditoriums; nor are these expectations disappointed when the phenomenon is submitted to exact measurement. It was not easy to find satisfactory conditions for the experiment, for not many rooms are now constructed in which plaster on studding, and sufficiently thin, forms a very considerable factor. After long waiting a room suitable in every respect, except location, became available. Its floor, its whole wall, indeed, its ceiling was of pine sheathing. The only other material entering into its construction was glass in the two windows and in the door. Unfortunately, the room was on a prominent street, and immediately adjacent was an all-night lunch room. Accurate experiments were out of the question while the lunch room was in use, and it was, therefore, bought out and closed for a few nights. Even with the freedom from noise thus secured, the experiments were not totally undisturbed. The traffic past the building did not stop sufficiently to permit of any observations until after two o'clock in the morning, and began again by four. During the intervening two hours, it was possible to snatch periods for observation, but even these periods were disturbed through the curiosity of passers and the more legitimate concern of the police.

Anticipating the phenomenon of resonance in wood in a more marked degree than in any other material, new apparatus was designed permitting of measurements at more frequent intervals of pitch. The new apparatus was not available when the work began and the coefficients for the wood were determined at octave intervals, with results as follows:

C_1, .064; C_2, .098; C_3, .112; C_4, .104; C_5, .081; C_6, .082; C_7, .113.

These results when plotted showed clearly a very marked resonance. The more elaborate apparatus was hastened to completion and the coefficients of absorption determined for the intermediate notes of E and G in each of the middle four octaves. The results of both sets of experiments when plotted together give Curve 5 in Fig. 1. The accuracy with which these fourteen points fall on a smooth curve drawn through them is all that could be expected in view of the conditions under which the experiment was conducted and the limited time available. Only one point falls far from the curve, that for middle C (C_3, 256). The general trend of the curve, however, is established beyond reasonable doubt. It is interesting to note the very great differences between this curve and those obtained for solid walls, and even for plastered walls. It is especially interesting to note the great absorption due to the resonance between the natural vibration of the walls and the sound, and to observe that this maximum point of resonance lies in the lower part, although not in the lowest part, of the range of pitch tested. The pitch of this resonance is determined by the nature of the wood, its thickness, and the distance apart of the studding on which it is supported. The wood tested was North Carolina pine, five-eighths of an inch in thickness and on studding fourteen inches apart. It is, perhaps, not superfluous to add at this time that a denser wood would have had a lower pitch for maximum resonance, other conditions being alike; an increased thickness would have raised the pitch of the resonance; while an increased distance between the studding would have lowered it. Finally it should be added that the best acoustical condition both for music and for speaking would have been with the maximum resonance an octave above rather than at middle C.

Even more interesting is the study of ceramic tile made at the request of Messrs. Cram, Goodhue, and Ferguson. The investigation had for its first object the determination of the acoustical value of the tile as employed in the groined arches of the Chapel of the United States Military Academy at West Point. The investigation then widened its scope, and, through the skill and great knowledge of ceramic processes of Mr. Raphael Guastavino, led to really remarkable results in the way of improved acoustical efficiency. The resulting construction has not only been approved by architects as equal, if not better, in architectural appearance to ordinary tile construction, but it is, so far as the writer knows, the first finished structural surface of large acoustical efficiency. Its random use does not, of course, guarantee good acoustical quality in an auditorium, for that depends on the amount used and the surface covered.

The first investigation was in regard to tile used at West Point, with the following result:

C_1, .012; C_2, .013; C_3, .018; C_4, .029; C_5, .040; C_6, .048; C_7, .053.

These are plotted in Curve 1, Fig. 2. The first endeavors to improve the tile acoustically had very slight results, but such as they were they were incorporated in the tile of the ceiling of the First Baptist Church in Pittsburgh (Curve 2, Fig. 2).

C_1, .028; C_2, .030; C_3, .038; C_4, .053; C_5, .080; C_6, .102; C_7, .114.

There was no expectation that the results of this would be more than a very slight amelioration of the difficulties which were to be expected in the church. In consequence of its use, the tile may be distinguished for purposes of tabulation as Pittsburgh Tile. Without following the intermediate steps, it is sufficient to say that the experiments were continued nearly two years longer and ultimately led to a tile which for the conveniences of tabulation we will call Acoustical Tile. The resulting absorbent power is far beyond what was conceived to be possible at the beginning of the investigation, and makes the construction in which this tile is incorporated unique in acoustical value among rigid structures. The coefficients for this construction are as follows:

C_1, .064; C_2, .068; C_3, .117; C_4, .188; C_5, .250; C_6, .258; C_7, .223,

graphically shown in Curve 3, Fig. 2. It is not a panacea. There is, on the other hand, no question but that properly used it will very greatly ameliorate the acoustical difficulties when its employment is practicable, and used in proper locations and amounts will render the acoustics of many auditoriums excellent which would otherwise be intolerable. It has over sixfold the absorbing power of any existing masonry construction and one-third the absorbing power of the

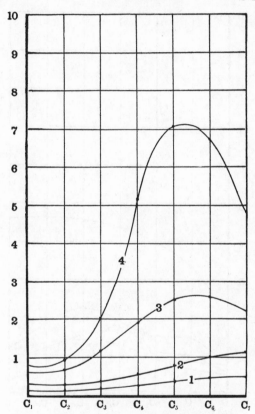

Fig. 2. Absorbing power: 1, West Point tile; 2, Pittsburgh tile; 3, acoustical tile; 4, best felt.

best known felt plotted on the same diagram for comparison (Curve 4). It is a new factor at the disposal of the architect.

Chairs and Audience

Equally important with the wall and ceiling-surfaces of an auditorium are its contents, especially the seats and the audience.

In expressing the coefficients of absorption for objects which are themselves units and which cannot be figured as areas, the coefficients depend on the system of measurement employed, Metric or English. While the international or metric system has become universal except in English speaking countries, and even in England and America in many fields, it has not yet been adopted by the

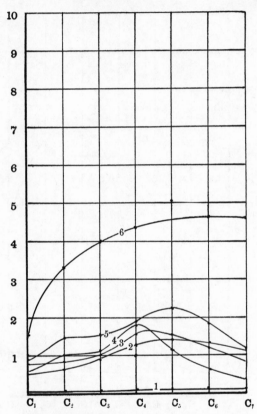

Fig. 3. Absorbing power: 1, bent wood chairs; 2, 3, 4, and 5, various kinds of pew cushions as described in text; 6, audience per person.

architectural profession and by the building trades, and therefore these coefficients will be given in both systems.

Ash settees or chairs, such as are ordinarily to be found in a college lecture-room, have exceedingly small absorbing powers. Such furniture forms a very small factor in the acoustics of any auditorium in which it is employed. The coefficients for ash chairs are as follows (Curve 1, Fig. 3):

Metric

C_1, .014; C_2, .014; C_3, .015; C_4, .016; C_5, .017; C_6, .019; C_7, .021.

English

C_1, .15; C_2, .15; C_3, .16; C_4, .17; C_5, .18; C_6, .20; C_7, 23.

The coefficients for settees were also determined, but differ so little from those for chairs that this paper will not be burdened with them. When, however, the seats are upholstered, they immediately become a considerable factor in the acoustics of an empty, or partially empty, auditorium. Of course the chairs either upholstered or unupholstered are not a factor in the acoustics of the auditorium when occupied. The absorbing power of cushions depends in considerable measure upon the nature of the covering and upon the nature of the padding. The cushions experimented upon were such as are employed in church pews, but the coefficients are expressed in terms of the cushion which would cover a single seat. The coefficients are as follows:

Cushions of wiry vegetable fiber covered with canvas and a thin damask cloth (Curve 2, Fig. 3):

Metric

C_1, .060; C_2, .070; C_3, .097; C_4, .135; C_5, .148; C_6, .132; C_7, .115.

English

C_1, .64; C_2, .75; C_3, 1.04; C_4, 1.45; C_5, 1.59; C_6, 1.42; C_7, 1.24.

Cushions of long hair covered with canvas and with an outer covering of plush (Curve 3, Fig. 3):

Metric

C_1, .080; C_2, .092; C_3, .105; C_4, .165; C_5, .155; C_6, .128; C_7, .085.

English

C_1, .86; C_2, .99; C_3, 1.13; C_4, 1.77; C_5, 1.67; C_6, 1.37; C_7, .91.

Cushions of hair covered with canvas and an outer covering of thin leatherette (Curve 4, Fig. 3):

Metric

C_1, .062; C_2, .105; C_3, .118; C_4, .180; C_5, .118; C_6, .068; C_7, .040.

English

C_1, .67; C_2, 1.13; C_3, 1.27; C_4, 1.93; C_5, 1.27; C_6, .73; C_7, .43.

Elastic felt cushions of commerce, elastic cotton covered with canvas and a short nap plush (Curve 5, Fig. 3):

Metric

C_1, .092; C_2, .155; C_3, .175; C_4, .190; C_5, .258; C_6, .182; C_7, 120.

English

C_1, .99; C_2, 1.66; C_3, 1.88; C_4, 2.04; C_5, 2.77; C_6, 1.95; C_7, 1.29.

Of all the coefficients of absorption, obviously the most difficult to determine are those for the audience itself. It would not at all serve to experiment on single persons and to assume that when a number are seated together, side by side, and in front of one another, the absorbing power is the same. It is necessary to make the experiment on a full audience, and to conduct such an experiment requires the nearly perfect silence of several hundred persons, the least noise on the part of one vitiating the observation. That the experiment was ultimately successful beyond all expectation is due to the remarkable silence maintained by a large Cambridge audience that volunteered itself for the purpose, not merely once, but on four separate occasions. The coefficients of absorption thus determined lie, with but a single exception, on a smooth curve (Curve 6, Fig. 3). The single exception was occasioned by the sound of a distant street car. Correcting this observation to the curve, the coefficients for an audience per person are as follows:

Metric

C_1, .160; C_2, .332; C_3, .395; C_4, .440; C_5, .455; C_6, .460; C_7, .460.

English

C_1, 1.72; C_2, 3.56; C_3, 4.25; C_4, 4.72; C_5, 4.70; C_6, 4.95; C_7, 4.95.

Fabrics

It is evident from the above discussion that fabrics are high absorbents of sound. How effective any particular fabric may be, depends not merely on the texture of its surface and the material, but upon the weave or felting throughout its body, and of course, also upon its thickness. An illuminating study of this question can be made by means of the curves in Fig. 4. In this figure are plotted the coefficients of absorption for varying thicknesses of felt. Curve 1 is the absorption curve for felt of one-half inch thickness.

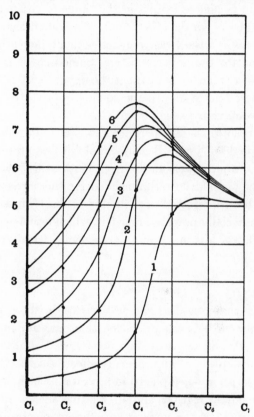

Fig. 4. Absorbing power of felt of varying thickness, from one-half to three inches, showing by extrapolation the absorption by thin fabrics of the upper register only.

Curve 2 of felt of one inch thickness, and so on up to Curve 6, which is for felt of three inches in thickness. It is interesting to contemplate what the result of the process would be were it continued to greater thickness, or in the opposite direction to felt of less and less thickness. It is inconceivable that felt should be used more than three inches in thickness and, therefore, extrapolation in this direction is of academic interest only. On the other hand, felt with decreasing thickness corresponds more and more to ordinary fabrics. If this process were carried to an extreme, it would show the effect of cheesecloth or bunting as a factor in the acoustics of an auditorium. It is obvious that very thin fabrics absorb only the highest notes and are negligible factors in the range of either the speaking voice or of music. On the other hand, it is evident that great thickness of felt absorbs the lower register without increasing whatever its absorption for the upper register. Sometimes it is desirable to absorb the lower register, sometimes the upper register, but far more often it is desirable to absorb the sounds from C_3 to C_6, but especially in the octave between C_4 and C_5.

The felt used in these experiments was of a durable nature and largely composed of jute. Because wool felt and ordinary hair felt are subject to rapid deterioration from moths, this jute felt was the only one which could be recommended for the correction of auditoriums until an interested participator in these investigations developed an especially prepared hair felt, which is less expensive than jute felt, but which is much more absorbent. Its absorption curve is plotted in Fig. 2.

LOCATION

Such a discussion as this should not close without pointing out the triple relation between pitch, location, and apparent power of absorption. This is shown in Fig. 5. Curve 1 shows the true coefficient of absorption of an especially effective felt. Curve 2 is its apparent absorption when placed in a position which is one of loudness for the lower register and of relative silence for the upper register. Curve 3 is the apparent coefficient of absorption of the same felt when placed in a position in the room of maximum loudness for all registers. It is evident from these three curves that in one position a felt may lose thirty per cent and over of its efficiency in the most significant register, or may have its efficiency nearly doubled. These curves relate to the efficiency of the felt in its effect on general reverberation. Its efficiency in the reduction of a discrete echo is dependent to an even greater degree on its location than on pitch.

The above are the coefficients of absorption for most materials usually occurring in auditorium construction, but there are certain omissions which it is highly desirable to supply, particularly notice-able among these is the absorption curve for glass and for old plaster.

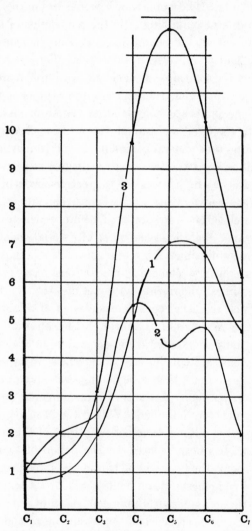

Fig. 5. Double dependence of absorbing power on pitch and on location, showing one of the sources of error which must be guarded against in the determination of coefficients of absorption and in the use of absorbing materials.

It is necessary for such experiments that rooms practically free from furniture should be available and that the walls and ceiling of the room should be composed in a large measure of the material to be tested. The author would appreciate any opportunity to carry out such experiments. The opportunity would ordinarily occur in the construction of a new building or in the remodeling of an old one.

It may be not wholly out of place to point out another modern acoustical difficulty and to seek opportunities for securing the necessary data for its solution. Coincident with the increased use of reënforced concrete construction and some other building forms there has come increased complaint of the transmission of sound from room to room, either through the walls or through the floors. Whether the present general complaint is due to new materials and new methods of construction, or to a greater sensitiveness to unnecessary noise, or whether it is due to greater sources of disturbance, heavier traffic, heavier cars and wagons, elevators, and elevator doors, where elevators were not used before, — whatever the cause of the annoyance there is urgent need of its abatement in so far as it is structurally possible. Moreover, several buildings have shown that not infrequently elaborate precautions have resulted disastrously, sometimes fundamentally, sometimes through the oversight of details which to casual consideration seem of minor importance. Here, as in the acoustics of auditoriums, the conditions are so complicated that only a systematic and accurately quantitative investigation will yield safe conclusions. Some headway, perhaps half a year's work, little more than a beginning, was made in this investigation some years ago. Methods of measurements were developed and some results were obtained. Within the past month the use of a room in a new building, together with that of the room immediately below it, has been secured for the period of two years. Between these rooms the floor will be laid in reënforced concrete of two thicknesses, five inches and ten inches, in hollow tile, in brick arch, in mill construction, and with hung ceiling, and the transmission of sound tested in each case. The upper surface of the floor will be laid in tile, in hardwood, with and without sound-deadening lining, and covered with linoleum and cork, and its noise to the tread measured.

However, such experiments but lay the foundation. What is needed are tests of the walls and floors of rooms of various sizes, and of the more varied construction which occurs in practice, in rooms connecting with offsets and different floor levels, — the complicated condition of actual building as against the simplified conditions of an orderly experiment. The one will give numerical coefficients, the other, if in sufficiently full measure, will give experience leading to generalization which may be so formulated as to be of wide value. What is therefore sought is the opportunity to experiment in rooms of varied but accurately known construction, especially where the insulation has been successful. Unfortunately, with modern building materials acoustical difficulties of all sorts are very numerous.

9

ARCHITECTURAL ACOUSTICS[1]

Because familiarity with the phenomena of sound has so far out-stripped the adequate study of the problems involved, many of them have been popularly shrouded in a wholly unnecessary mystery. Of none, perhaps, is this more true than of architectural acoustics. The conditions surrounding the transmission of speech in an en-closed auditorium are complicated, it is true, but are only such as will yield an exact solution in the light of adequate data. It is, in other words, a rational engineering problem.

The problem of architectural acoustics is necessarily complex, and each room presents many conditions which contribute to the result in a greater or less degree, according to circumstances. To take justly into account these varied conditions, the solution of the problem should be quantitative, not merely qualitative; and to reach its highest usefulness and the dignity of an engineering science it should be such that its application can precede, not merely follow, the construction of the building.

In order that hearing may be good in any auditorium it is neces-sary that the sound should be sufficiently loud, that the simulta-neous components of a complex sound should maintain their proper relative intensities, and that the successive sounds in rapidly moving articulation, either of speech or of music, should be clear and distinct, free from each other and from extraneous noises. These three are the necessary, as they are the entirely sufficient, conditions for good hearing. Scientifically the problem involves three factors: rever-beration, interference, and resonance. As an engineering problem it involves the shape of the auditorium, its dimensions, and the materials of which it is composed.

Sound, being energy, once produced in a confined space, will continue until it is either transmitted by the boundary walls or is transformed into some other kind of energy, generally heat. This process of decay is called absorption. Thus, in the lecture-room of

[1] The Journal of the Franklin Institute, January, 1915.

Harvard University, in which, and in behalf of which, this investigation was begun, the rate of absorption was so small that a word spoken in an ordinary tone of voice was audible for five and a half seconds afterwards. During this time even a very deliberate speaker would have uttered the twelve or fifteen succeeding syllables. Thus the successive enunciations blended into a loud sound, through which and above which it was necessary to hear and distinguish the orderly progression of the speech. Across the room this could not be done; even near the speaker it could be done only with an effort wearisome in the extreme if long maintained. With an audience filling the room the conditions were not so bad, but still not tolerable. This may be regarded, if one so chooses, as a process of multiple reflection from walls, from ceiling, and from floor, first from one and then another, losing a little at each reflection until ultimately inaudible. This phenomenon will be called reverberation, including, as a special case, the echo. It must be observed, however, that, in general, reverberation results in a mass of sound filling the whole room and incapable of analysis into its distinct reflections. It is thus more difficult to recognize and impossible to locate. The term "echo" will be reserved for that particular case in which a short, sharp sound is distinctly repeated by reflection, either once from a single surface, or several times from two or more surfaces. In the general case of reverberation we are concerned only with the rate of decay of the sound. In the special case of the echo we are concerned not merely with its intensity, but with the interval of time elapsing between the initial sound and the moment it reaches the observer. In the room mentioned as the occasion of this investigation no discrete echo was distinctly perceptible, and the case will serve excellently as an illustration of the more general type of reverberation. After preliminary gropings, first in the literature and then with several optical devices for measuring the intensity of sound, all established methods were abandoned. Instead, the rate of decay was measured by measuring what was inversely proportional to it, — the duration of audibility of the reverberation, or, as it will be called here, the duration of audibility of the residual sound. These experiments may be explained to advantage here, for they will give more clearly than would abstract discussion an idea of the nature of reverberation. Broadly considered, there are two, and only two, variables in a room, — shape (including size) and materials (including furnishings). In designing an auditorium an architect can give consideration to both; in repair work for bad acoustic conditions it is generally impracticable to change the shape, and only variations in materials and furnishings are allowable. This was, therefore, the line of work in this case. It was evident that, other things being equal, the rate at which the reverberation would disappear was proportional to the rate at which the sound was absorbed. The first work, therefore, was to determine the relative absorbing power

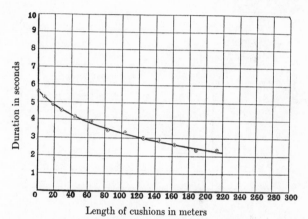

Fig. 1. Curve showing the relation of the duration of the residual sound to the added absorbing material.

of various substances. With an organ pipe as a constant source of sound, and a suitable chronograph for recording, the duration of audibility of a sound after the source had ceased in this room when empty was found to be 5.62 seconds. All the cushions from the seats in Sanders Theatre were then brought over and stored in the lobby. On bringing into the lecture-room a number of cushions, having a total length of 8.2 meters, the duration of audibility fell to 5.33 seconds. On bringing in 17 meters the sound in the room after the organ pipe ceased was audible for but 4.94 seconds. Evidently the cushions were strong absorbents and rapidly improving the room, at least to the extent of diminishing the reverberation. The result was interesting and the process was continued. Little by little the cushions were brought into the room, and each time the duration of audibility was measured. When all the seats (436 in number) were covered, the sound was audible for 2.03 seconds. Then the aisles were covered, and then the platform. Still there were more cushions, — almost half as many more. These were brought into the room, a few at a time, as before, and draped on a scaffolding that had been erected around the room, the duration of the sound being recorded each time. Finally, when all the cushions from a theatre seating nearly fifteen hundred persons were placed in the room — covering the seats, the aisles, the platform, the rear wall to the ceiling — the duration of audibility of the residual sound

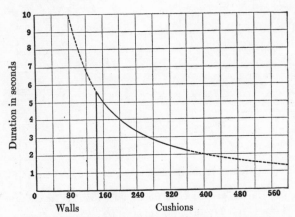

Fig. 2. Curve 5 plotted as part of its corresponding rectangular hyperbola. The solid part was determined experimentally; the displacement of this to the right measures the absorbing power of the walls of the room.

was 1.14 seconds. This experiment, requiring, of course, several nights' work, having been completed, all the cushions were removed and the room was in readiness for the test of other absorbents. It was evident that a standard of comparison had been established. Curtains of chenille, 1.1 meters wide and 17 meters in total length, were draped in the room. The duration of audibility was then 4.51 seconds. Turning to the data that had just been collected, it appeared that this amount of chenille was equivalent to 30 meters of Sanders Theatre cushions. Oriental rugs (Herez, Demirjik, and Hindoostanee) were tested in a similar manner, as were also cretonne cloth, canvas, and hair felt. Similar experiments, but in a smaller room, determined the absorbing power of a man and of a woman, always by determining the number of running meters of Sanders Theatre cushions that would produce the same effect. This process of comparing two absorbents by actually substituting one for the other is laborious, and it is given here only to show the first steps in the development of a method. Without going into details, it is sufficient here to say that this method was so perfected as to give not merely relative, but absolute, coefficients of absorption.

In this manner a number of coefficients of absorption were determined for objects and materials which could be brought into and removed from the room, for sounds having a pitch an octave above middle C. In the following table the numerical values are the absolute coefficients of the absorption:

Oil paintings, inclusive of frames	.28
Carpet rugs	.20
Oriental rugs, extra heavy	.29
Cheesecloth	.019
Cretonne cloth	.15
Shelia curtains	.23
Hair felt, 2.5 cm. thick, 8 cm. from wall	.78
Cork, 2.5 cm. thick, loose on floor	.16
Linoleum, loose on floor	.12

When the objects are not extended surfaces, such as carpets or rugs, but essentially spacial units, it is not easy to express the absorption as an absolute coefficient. In the following table the absorption of each object is expressed in terms of a square meter of complete absorption:

Audience, per person	.44
Isolated woman	.54
Isolated man	.48
Plain ash settees	.039
Plain ash settees, per single seat	.0077
Plain ash chairs, "bent wood"	.0082
Upholstered settees, hair and leather	1.10
Upholstered settees, per single seat	.28
Upholstered chairs similar in style	.30
Hair cushions, per seat	.21
Elastic felt cushions, per seat	.20

Of even greater importance was the determination of the coefficient of absorption of floors, ceilings, and wall-surfaces. The accomplishment of this called for a very considerable extension of the method adopted. If the reverberation in a room as changed by the addition of absorbing material be plotted, the resulting curve will be found to be a portion of an hyperbola with displaced axes. An example of such a curve, as obtained in the lecture-room of the Fogg Art Museum, in Cambridge, is plotted in the diagram, Fig. 1. If now the origin of this curve be displaced so that the axes of coördinates are the asymptotes of the rectangular hyperbola, the displacement of the origin measures the initial ab-

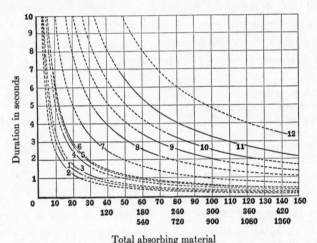

Fig. 3. The curves of Figs. 8 and 9 entered as parts of their corresponding rectangular hyperbolas. Three scales are employed for the volumes, by groups 1–7, 8–11, and 12.

sorbing power of the room, its floors, walls, and ceilings. Such experiments were carried out in a large number of rooms in which the different component materials entered in very different degrees, and an elimination between these different experiments gave the following coefficient of absorption for different materials:

Open window....................................	1.000
Wood sheathing (hard pine).....................	.061
Plaster on wood lath...........................	.034
Plaster on wire lath...........................	.033
Glass, single thickness........................	.027
Plaster on tile................................	.025
Brick set in Portland cement...................	.025

If the experiments in these rooms are plotted in a single diagram, the result is a family of hyperbolae showing a very interesting relationship to the volumes of the rooms. Indeed, if from these hyperbolas the parameter, which equals the product of the coordinates, be determined, it will be found to be linearly proportional to the volume of the room. These results are plotted in Fig. 4, showing how strict the proportionality is even over a very great range in volume. We have thus at hand a ready method of

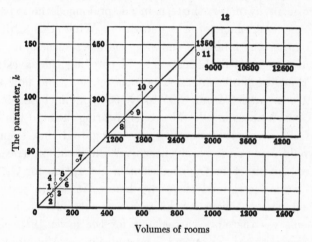

Fig. 4. The parameter, k, plotted against the volumes of the rooms, showing the two proportional.

calculating the reverberation for any room, its volume and the materials of which it is composed being known.

The first five years of the investigation were devoted to violin C, the C an octave above middle C, having a vibration frequency of 512 vibrations per second. This pitch was chosen because, in the art of telephony, it was regarded at that time as the characteristic pitch determining the conditions of articulate speech. The planning of Symphony Hall in Boston forced an extension of this investigation to notes over the whole range of the musical scale, three octaves below and three octaves above violin C.

In the very nature of the problem, the most important datum is the absorption coefficient of an audience, and the determination of this was the first task undertaken. By means of a lecture on one of the recent developments of physics, wireless telegraphy, an audience was thus drawn together and at the end of the lecture requested to remain for the experiment. In this attempt the effort was made to determine the coefficients for the five octaves from C_2128 to C_62048, including notes E and G in each octave. For several reasons the experiment was not a success. A threatening thunderstorm made the audience a small one, and the sultriness of the atmosphere made open windows necessary, while the attempt to cover so many notes, thirteen in all, prolonged the experiment beyond the endurance of the audience. While this experiment failed, another the following summer was more successful. In the year that had elapsed the necessity of carrying the investigation further than the limits intended became evident, and now the experiment was carried from C_164 to C_74096, but included only the C notes, seven notes in all. Moreover, bearing in mind the experiences of the previous summer, it was recognized that even seven notes would come dangerously near overtaxing the patience of the audience. Inasmuch as the coefficient of absorption for C_4512 had already been determined six years before, in the investigations mentioned, the coefficient for this note was not redetermined. The experiment was therefore carried out for the lower three and the upper three notes of the seven. The audience, on the night of this experiment, was much larger than that which came the previous summer, the night was a more comfortable one, and it was possible to close the windows during the experiment. The conditions were thus fairly satisfactory. In order to get as much data as possible, and in as short a time, there were nine observers stationed at different points in the room. These observers, whose kindness and skill it is a pleasure to acknowledge, had prepared themselves, by previous practice, for this one experiment. The results of the experiment are shown on the lower curve in Fig. 5. This curve gives the coefficient of absorption per person. It is to be observed that one of the points falls clearly off the smooth curve drawn through the other points.[1] The observations on which this point is based were, however, much disturbed by a street car passing not far from the building, and the departure of this observation from the curve does not

---

[1] This point, evidently on the ordinate C_5, is omitted in the original cut. — Editor.

indicate a real departure in the coefficient, nor should it cast much doubt on the rest of the work, in view of the circumstances under which it was secured. Counteracting the, perhaps, bad impression

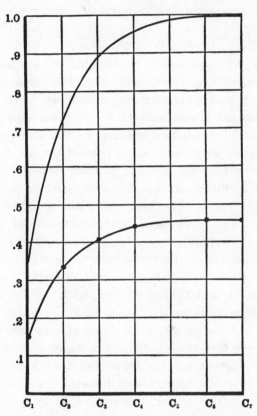

FIG. 5. The absorbing power of an audience for different notes. The lower curve represents the absorbing power of an audience per person. The upper curve represents the absorbing power of an audience per square meter as ordinarily seated. The vertical ordinates are expressed in terms of total absorption by a square meter of surface. For the upper curve the ordinates are thus the ordinary coefficients of absorption. The several notes are at octave intervals, as follows: $C_1 64$, $C_2 128$, C_3 (middle C) 256, $C_4 512$, $C_5 1024$, $C_6 2048$, $C_7 4096$.

which this point may give, it is a considerable satisfaction to note how accurately the point for $C_4 512$, determined six years before by a different set of observers, falls on the smooth curve through the remaining points. In the audience on which these observations were taken there were 77 women and 105 men. The courtesy of the audience in remaining for the experiment and the really remarkable silence which they maintained are gratefully acknowledged.

The next experiment was on the determination of the absorption of sound by wood sheathing. It is not an easy matter to find conditions suitable for this experiment. The room in which the absorption by wood sheathing was determined in the earlier experiments was not available for these. It was available then only because the building was new and empty. When these more elaborate experiments were under way the room became occupied, and in a manner that did not admit of its being cleared. Quite a little searching in the neighborhood of Boston failed to discover an entirely suitable room. The best one available adjoined a night lunch room. The night lunch was bought out for a couple of nights, and the experiment was tried. The work of both nights was much disturbed. The traffic past the building did not stop until nearly two o'clock, and began again at four. The interest of those passing on foot throughout the night, and the necessity of repeated explanations to the police, greatly interfered with the work. This detailed statement of the conditions under which the experiment was tried is made by way of explanation of the irregularity of the observations recorded on the curve, and of the failure to carry this particular line of work further. The first night seven points were obtained for the seven notes $C_1 64$ to $C_7 4096$. The reduction of these results on the following day showed variations indicative of maxima and minima, which, to be accurately located, would require the determination of intermediate points. In the experiment the following night points were determined for the E and G notes in each octave between $C_2 128$ and $C_6 2048$. Other points would have been determined, but time did not permit. It is obvious that the intermediate points in the lower and in the higher octave were desirable, but no pipes were to be had on such short notice for this part of the range, and in their absence the data could not be obtained. In the diagram, Fig. 6, the points lying on the vertical lines were determined the first night. The points lying between the vertical lines were determined the second night. The accuracy with which these points fall on a smooth curve is, perhaps,

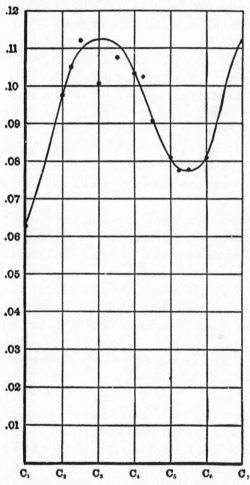

FIG. 6. The absorbing power of wood sheathing, two centimeters thick, North Carolina pine. The observations were made under very unsuitable conditions. The absorption is here due almost wholly to yielding of the sheathing as a whole, the surface being shellacked, smooth, and non-porous. The curve shows one point of resonance within the range tested, and the probability of another point of resonance above. It is not possible now to learn as much in regard to the framing and arrangement of the studding in the particular room tested as is desirable. C_3 (middle C) 256.

all that could be expected in view of the difficulty under which the observations were conducted and the limited time available. One point in particular falls far off from this curve, the point for C_3256, by an amount which is, to say the least, serious, and which can be justified only by the conditions under which the work was done. The general trend of the curve seems, however, established beyond reasonable doubt. It is interesting to note that there is one point of maximum absorption, which is due to resonance between the walls and the sound, and that this point of maximum absorption lies in the lower part, though not in the lowest part, of the range of pitch tested. It would have been interesting to determine, had the time and facilities permitted, the shape of the curve beyond C_74096, and to see if it rises indefinitely, or shows, as is far more likely, a succession of maxima.

The experiment was then directed to the determination of the absorption of sound by cushions, and for this purpose return was made to the constant-temperature room. Working in the manner indicated in the earlier papers for substances which could be carried in and out of a room, the curves represented in Fig. 7 were obtained. Curve 1 shows the absorption coefficient for the Sanders Theatre cushions, with which the whole investigation was begun ten years ago. These cushions were of a particularly open grade of packing, a sort of wiry grass or vegetable fiber. They were covered with canvas ticking, and that, in turn, with a very thin cloth covering. Curve 2 is for cushions borrowed from the Phillips Brooks House. They were of a high grade, filled with long, curly hair, and covered with canvas ticking, which was, in turn, covered by a long nap plush. Curve 3 is for the cushions of Appleton Chapel, hair covered with a leatherette, and showing a sharper maximum and a more rapid diminution in absorption for the higher frequencies, as would be expected under such conditions. Curve 4 is probably the most interesting, because for more standard commercial conditions ordinarily used in churches. It is to be observed that all four curves fall off for the higher frequencies, all show a maximum located within an octave, and three of the curves show a curious hump in the second octave. This break in the curve is a genuine phenomenon, as it was tested time after time. It is perhaps due to a secondary resonance, and it is to be observed that it is the more pronounced in those curves that have the sharper resonance in their principal maxima.

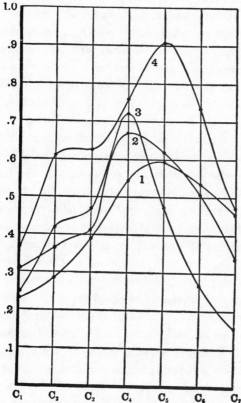

Fig. 7. The absorbing power of cushions. Curve 1 is for "Sanders Theatre" cushions of wiry vegetable fiber, covered with canvas ticking and a thin cloth. Curve 2 is for "Brooks House" cushions of long hair, covered with the same kind of ticking and plush. Curve 3 is for "Appleton Chapel" cushions of hair, covered with ticking and a thin leatherette. Curve 4 is for the elastic felt cushions of commerce, of elastic cotton, covered with ticking and short nap plush. The absorbing power is per square meter of surface. C_3 (middle C) 256.

In both articulate speech and in music the source of sound is rapidly and, in general, abruptly changing in pitch, quality, and loudness. In music one pitch is held during the length of a note.

In articulate speech the unit or element of constancy is the syllable. Indeed, in speech it is even less than the length of a syllable, for the open vowel sound which forms the body of a syllable usually has a consonantal opening and closing. During the constancy of an element, either of music or of speech, a train of sound-waves spreads spherically from the source, just as a train of circular waves spreads outward from a rocking boat on the surface of still water. Different portions of this train of spherical waves strike different surfaces of the auditorium and are reflected. After such reflection they begin to cross each other's paths. If their paths are so different in length that one train of waves has entirely passed before the other arrives at a particular point, the only phenomenon at that point is prolongation of the sound. If the space between the two trains of waves be sufficiently great, the effect will be that of an echo. If there be a number of such trains of waves thus widely spaced, the effect will be that of multiple echoes. On the other hand, if two trains of waves have traveled so nearly equal paths that they overlap, they will, dependent on the difference in length of the paths which they had traveled, either reënforce or mutually destroy each other. Just as two equal trains of water-waves crossing each other may entirely neutralize each other if the crest of one and the trough of the other arrive together, so two sounds, coming from the same source, in crossing each other may produce silence. This phenomenon is called interference, and is a common phenomenon in all types of wave-motion. Of course, this phenomenon has its complement. If the two trains of water-waves so cross that the crest of one coincides with the crest of the other and trough with trough, the effects will be added together. If the two sound-waves be similarly retarded, the one on the other, their effects will also be added. If the two trains of waves be equal in intensity, the combined intensity will be quadruple that of either of the trains separately, as above explained, or zero, depending on their relative retardation. The effect of this phenomenon is to produce regions in an auditorium of loudness and regions of comparative or even complete silence. It is a partial explanation of the so-called deaf regions in an auditorium.

It is not difficult to observe this phenomenon directly. It is difficult, however, to measure and record the phenomenon in such a manner as to permit of an accurate chart of the result. Without going into the details of the method employed, the result of these

Fig. 8. Distribution of intensity on the head level in a room with a barrel-shaped ceiling, with center of curvature on the floor level.

measurements for a room very similar to the Congregational Church in Naugatuck, Connecticut, is shown in the accompanying chart. The room experimented in was a simple, rectangular room with plain side walls and ends and with a barrel or cylindrical ceiling. The result is clearly represented in Fig. 8, in which the intensity of the sound has been indicated by contour lines in the manner employed in the drawing of the geodetic survey maps. The phenomenon indicated in these diagrams was not ephemeral, but was constant so long as the source of sound continued, and repeated itself with almost perfect accuracy day after day. Nor was the phenom-

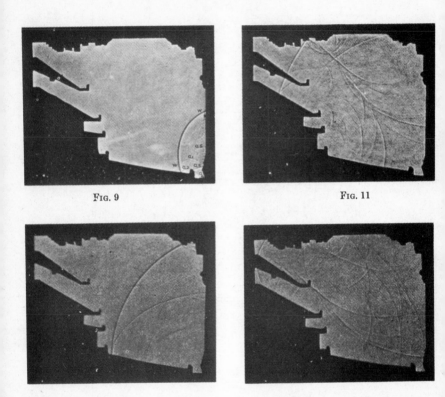

Fig. 9

Fig. 11

Fig. 10

Fig. 12

enon one which could be observed merely instrumentally. To an observer moving about in the room it was quite as striking a phenomenon as the diagrams suggest. At the points in the room indicated as high maxima of intensity in the diagram the sound was so loud as to be disagreeable, at other points so low as to be scarcely audible. It should be added that this distribution of intensity is with the source of sound at the center of the room. Had the source of sound been at one end and on the axis of the cylindrical ceiling, the distribution of intensity would still have been bilaterally symmetrical, but not symmetrical about the transverse axis.

When a source of sound is maintained constant for a sufficiently long time — a few seconds will ordinarily suffice — the sound becomes steady at every point in the room. The distribution of the intensity

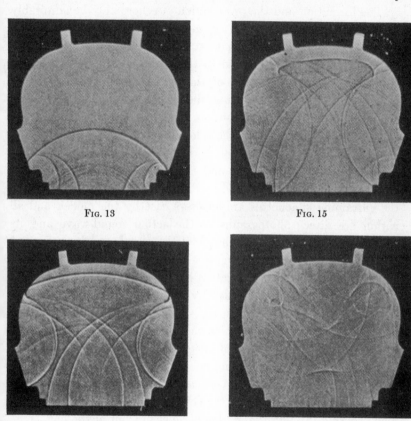

Fig. 13

Fig. 15

Fig. 14

Fig. 16

of sound under these conditions is called the interference system, for that particular note, of the room or space in question. If the source of sound is suddenly stopped, it requires some time for the sound in the room to be absorbed. This prolongation of sound after the source has ceased is called reverberation. If the source of sound, instead of being maintained, is short and sharp, it travels as a discrete wave or group of waves about the room, reflected from wall to wall, producing echoes. In the Greek theatre there was ordinarily but one echo, "doubling the case ending," while in the modern auditorium there are many, generally arriving at a less interval of time after the direct sound and therefore less distinguishable, but stronger and therefore more disturbing.

The formation and the propagation of echoes may be admirably studied by an adaptation of the so-called *schlieren-Methode* device for photographing air disturbances. It is sufficient here to say that the adaptation of this method to the problem in hand consists in the construction of a model of the auditorium to be studied to proper scale, and investigating the propagation through it of a proportionally scaled sound-wave. To examine the formation of echoes in a vertical section, the sides of a model are taken off and, as the sound is passing through it, it is illuminated instantaneously by the light from a very fine and somewhat distant electric spark. In the preceding illustrations, reduced from the photographs, the enframing silhouettes are shadows cast by the model, and all within are direct photographs of the actual sound-wave and its echoes. The four photographs show the sound and its echoes at different stages in their propagation through the room, the particular auditorium under investigation being the New Theatre in New York. It is not difficult to identify the master wave and the various echoes which it generates, nor, knowing the velocity of sound, to compute the interval at which the echo is heard.

To show the generation of echoes and their propagation in a horizontal plane, the ceiling and floor of the model are removed and the photograph taken in a vertical direction. The photographs shown in Figs. 13 to 16 show the echoes produced in the horizontal plane passing through the marble parapet in front of the box.

While these several factors, reverberation, interference, and echo, in an auditorium at all complicated are themselves complicated, nevertheless they are capable of an exact solution, or, at least, of a solution as accurate as are the architect's plans in actual construction. And it is entirely possible to calculate in advance of construction whether or not an auditorium will be good, and, if not, to determine the factors contributing to its poor acoustics and a method for their correction.

10

THE INSULATION OF SOUND [1]

THE insulation of sound as an unsolved problem in architectural acoustics was first brought to the writer's attention by the New England Conservatory of Music, immediately after its completion in 1904, and almost simultaneously in connection with a private house which had just been completed in New York. A few years later it was renewed by the Institute of Musical Art in New York. In the construction of all three buildings it had been regarded as particularly important that communication of sound from room to room should be avoided, and methods to that end had been employed which were in every way reasonable. The results showed that in this phase of architectural acoustics also there had not been a sufficiently searching and practical investigation and that there were no experimental data on which an architect could rely. As these buildings were the occasion for beginning this investigation, and were both instructive and suggestive, they are, with the consent of the architects, discussed here at some length.

The special method of construction employed in the New England Conservatory of Music was suggested to the architects by the Trustees of the Conservatory. The floor of each room was of semi-fireproof construction, cement between iron girders, on this a layer of plank, on this paper lining, and on top of this a floor of hard pine. Between each room for violin, piano, or vocal lessons was a compound wall, constructed of two partitions with an unobstructed air space between them. Each partition was of two-inch plaster block set upright, with the finishing plaster applied directly to the block. The walls surrounding the organ rooms were of three such partitions separated by two-inch air spaces. In each air space was a continuous layer of deadening cloth. The scheme was carried out consistently and with full regard to details, yet lessons conducted in adjacent rooms were disturbing to each other.

---

[1] The Brickbuilder, vol. xxiv, no. 2, February, 1915.

It is always easier to explain why a method does not work than to know in advance whether it will or will not. It is especially easy to explain why it does not work when not under the immediate necessity of correcting it or of supplying a better. This lighter rôle of the irresponsible critic was alone invited in the case of the New England Conservatory of Music, nor will more be ventured at the present moment.

There is no question whatever that the fundamental consideration on which the device hinged was a sound one. Any discontinuity diminishes the transmission of sound; and the transition from masonry to air is a discontinuity of an extreme degree. Two solid masonry walls entirely separated by an air space furnish a vastly better sound insulation than either wall alone. On the other hand, the problem takes on new aspects if a masonry wall be replaced by a series of screen walls, each light and flexible, even though they aggregate in massiveness the solid wall which they replace. Moreover, such screen walls can rarely be regarded as entirely insulated from each other. Granting that accidental communication has nowhere been established, through, for example, the extrusion of plaster, the walls are of necessity in communication at the floor, at the ceiling, at the sides, or at the door jambs; and the connection at the floor, at least, is almost certain to be good. Further, and of extreme importance, given any connection at all, the thinness of the screen walls renders them like drumheads and capable of large response to small excitation.

It may seem a remote parallel, but assume for discussion two buildings a quarter of a mile apart. With the windows closed, no ordinary sound in one building could be heard in the other. If, however, the buildings were connected by a single metal wire fastened to the centers of window panes, it would be possible not merely to hear from within one building to within the other, but with care to talk. On the other hand, had the wires been connected to the heavy masonry walls of the two buildings, such communication would have been impossible. This hypothetical case, though extreme, indeed perhaps the better because of its exaggeration, will serve to analyze the problem. Here, as in every case, the transmission of sound involves three steps, — the taking up of the vibration, the function of the nearer window pane, its transmission by the wire, and its communication to the air of the receiving room by the remote window. The three functions may be combined into one when a solid wall separates the two rooms, the taking up, transmitting, and emitting of the sound being scarcely separable processes. On the other hand, they are often clearly separable, as in the case of multiple screen walls.

In the case of a solid masonry wall, the transmission from surface to surface is almost perfect; but because of the great mass and rigidity of the wall, it takes up but little of the vibration of the incident sound. It is entirely possible to express by a not very complicated analytical equation the amount of sound which a wall of simple dimensions will take up and transmit in terms of the mass of the wall, its elasticity, and its viscosity, and the frequency of vibration of the sound. But such an equation, while of possible interest to physicists as an exercise, is of no interest whatever to architects because of the difficulty of determining the necessary coefficients.

In the case of multiple screen walls, the communication from wall to wall, through the intermediate air space or around the edges, is poor compared with the face to face communication of a solid wall. But the vibration of the screen wall exposed to the sound, the initial step in the process of transmission, is greatly enhanced by its light and flexible character. Similarly its counterpart, the screen wall, which by its vibration communicates the sound to the receiving room, is light, flexible, and responsive to relatively small forces. That this responsiveness of the walls compensates or more than compensates for the poor communication between them, is the probable explanation of the transmission between the rooms in the New England Conservatory.

The Institute of Musical Art in New York presented interesting variations of the problem. Here also the rooms on the second and third floors were intended for private instruction and were designed to be sound proof from each other, from the corridor, and from the rooms above and below. The walls separating the rooms from the corridors were double, having connection only at the door jambs and at the floor. The screen wall next the corridor was of terra cotta block, finished on the corridor side with plaster applied directly to the terra cotta. The wall next the room was of gypsum block, plastered and finished in burlap. In the air space between the two walls, deadening sheet was hung. The walls separating the rooms were of gypsum block and finished in hard plaster and burlap. As shown on the diagram (Fig. 1), these walls were cellular, one

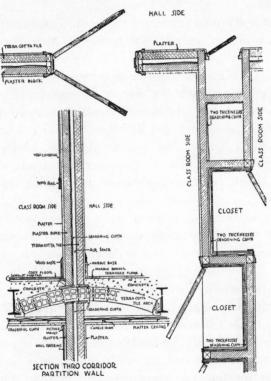

FIG. 1. Details of Construction, Institute of Musical Art, New York, N. Y.

of these cells being entirely enclosed in gypsum block, the others being closets opening the one to one room, the other to the other. The closets were lined with wood sheathing which was separated from the enclosing wall by a narrow space in which deadening sheet was hung in double thickness with overlapping joints. In the entirely enclosed cell, deadening sheet was also hung in double thickness.

It is not difficult to see, at least after the fact, why the deadening sheet in such positions was entirely without effect. The transverse masonry webs afforded a direct transmission from side to side of the compound wall that entirely overwhelmed the transmission through the air spaces. Had there been no necessity of closets, and therefore, no necessity of transverse web and had the two screen walls been truly insulated the one from the other, not merely over their area, but at the floor, at the ceiling, and at the edges, the insulation would have been much more nearly perfect.

The means which were taken to secure insulation at the base of the screen walls and to prevent the transmission of sound from floor to floor are exceedingly interesting. The floor construction consisted in hollow terra cotta tile arches, on top of this cinder concrete, on this sawdust mortar, and on the top of this cork flooring. Below the reënforced concrete arches were hung ceilings of plaster on wire lath. This hung ceiling was supported by crossed angle bars which were themselves supported by the I beams which supported the hollow terra cotta tile arches. In the air spaces between the tile arches and the hung ceilings, and resting on the latter, was deadening sheet. This compound floor of cork, sawdust mortar, cinder concrete, terra cotta tile, air space, and hung ceiling, with deadening sheet in the air spaces, has the air of finality, but was not successful in securing the desired insulation.

It is interesting to note also that the screen walls were separated from the floor arches on which they rested below and on which they abutted above by deadening sheet. It is possible that this afforded some insulation at the top of the wall, for the arch was not sustained by the wall, and the pressure at that point not great. At the bottom, however, it is improbable that the deadening sheet carried under the base offered an insulation of practical value. Under the weight of the wall it was probably compressed into a compact mass, whose rigidity was still further increased by the percolation through it of the cement from the surrounding concrete.

Finally, after the completion of the building, Mr. Damrosch, the director, had tried the experiment of covering the walls of one of the rooms to a depth of two inches with standard hair felt, with some, but almost negligible, effect on the transmission of sound.

Deadening sheet has been mentioned frequently. All indication of the special kind employed has been purposely omitted, for the discussion is concerned with the larger question of the manner of its use and not with the relative merits of the different makes.

The house in New York presented a problem even more interesting. It was practically a double house, one of the most imperative conditions of the building being the exclusion of sounds in the main part of the house from the part to the left of a great partition wall. This wall of solid masonry supported only one beam of the main house, was pierced by as few doors as possible — two — and by no steam or water pipes. The rooms were heated by independent fireplaces. The water pipes connected independently to the main. It had been regarded as of particular importance to exclude sounds from the two bedrooms on the second floor. The ceilings of the rooms below were, therefore, made of concrete arch; on top of this was spread three inches of sand, and on top of this three inches of lignolith blocks; on this was laid a hardwood floor; and finally, when the room was occupied, this floor was covered by very heavy and heavily padded carpets. From the complex floor thus constructed arose interior walls of plaster on wire lath on independent studding, supported only at the top where they were held from the masonry walls by iron brackets set in lignolith blocks. Each room was, therefore, practically a room within a room, separated below by three inches of sand and three inches of lignolith and on all sides and above by an air space. Notwithstanding this, the shutting of a door in any part of the main house could be heard, though faintly, in either bedroom. In the rear bedroom, from which the best results were expected, one could hear not merely the shutting of doors in the main part of the house, but the working of the feed pump, the raking of the furnace, and the coaling of the kitchen range. In the basement of the main dwelling was the servants' dining room. Rapping with the knuckles on the wall of this room produced in the bedroom, two stories up and on the other side of the great partition wall, a sound which, although hardly, as the architect expressed it, magnified, yet of astonishing loudness and clearness. In this case, the telephone-like nature of the process was even more clearly defined than in the other cases, for the distances concerned were much greater. The problem had many interesting aspects, but will best serve the present purpose if for the sake of simplicity and clearness it be held to but one, — the transmission of sound from the servants' dining room in the basement along the great eighteen-inch partition wall up two stories to the insulated bedroom above and opposite.

It is a fairly safe hazard that the sound on reaching the bedroom did not enter by way of the floor, for the combination of reënforced concrete, three inches of sand, three inches of lignolith block, and the wood flooring and carpet above, presented a combination of massive rigidity in the concrete arch, inertness in the sand and lignolith block, imperviousness in the hardwood floor, and absorption in the padded carpet which rendered insulation perfect, if perfect insulation be possible. No air ducts or steam or water pipes entered the room. The only conceivable communication, therefore, was through the walls or ceiling. The communication to the inner walls and ceiling from the surrounding structural walls was either through the air space or through the iron angle bars, which, set in lignolith blocks in the structural wall, retained erect and at proper distance the inner walls. Of the two means of communication, the air and the angle bars, the latter was probably the more important. It is interesting and pertinent to follow this line of communication, the masonry wall, the angle bars, and the screen walls, and to endeavor to discover if possible, or at least to speculate on the reason for its exceptional though unwelcome efficiency.

From the outset it is necessary to distinguish the transverse and the longitudinal transmission of sound in a building member, that is, to distinguish as somewhat different processes the transmission of sound from one room to an adjacent room through a separating wall or ceiling, from the transmission of sound along the floors from room to room, or along the vertical walls from floor to floor. Broadly, although the two are not entirely separable phenomena, one is largely concerned in the transmission of the sound of the voice, or the violin, or of other sources free from solid contact with the floor, and the other in the transmission of the sound of a piano or cello—instruments in direct communication with the building structure—or of noises involved in the operation of the building, dynamos, elevators, or the opening and closing of doors. In the building under consideration, the disturbing sounds were in every case communicated directly to the structure at a considerable distance and transmitted along the walls until ultimately communicated through the angle bars, if the angle bars were the means of communication, to the thin plaster walls which constituted the inner room. The special features thus emphasized were the longitudinal transmission of vibration by walls, floors, and structural beams, and the transformation of these longitudinal vibrations into the sound-producing transverse vibrations of walls and ceilings bounding the disturbed room. Many questions were raised which at the time could be only tentatively answered.

What manner of walls conduct the sound with the greater readiness? Is it true, as so often stated, that modern concrete construction has contributed to the recent prevalence of these difficulties? If so, is there a difference in this respect between stone, sand, and cinder concrete? In this particular building, the partition wall was of brick. Is there a difference due to the kind of brick employed, whether hard or soft? Or does the conduction of sound depend on the kind of mortar with which the masonry is set? If this seems trivial, consider the number of joints in even a moderate distance. Again, is it possible that sound may be transmitted along a wall without producing a transverse vibration, thus not entering the adjacent room? Is it possible that in the case of this private house had there been no interior screen wall the sound communicated to the room would have been less? We know that if the string of a string telephone passes through a room without touching, a conversation held over the line will be entirely inaudible in the room. Is it possible that something like this, but on a grand scale, may happen in a building? Or, again, is it possible that the iron brackets which connected the great partition wall to the screen wall magnified the motion and so the sound, as the lever on a phonograph magnifies its motion? These are not unworthy questions, even if ultimately the answer be negative.

The investigation divides itself into two parts, — the one dealing with partition walls especially constructed for the test, the other with existing structures wherever found in interesting form. The experiments of the former type were conducted in a special room, mentioned in some of the earlier papers (The Brickbuilder, January, 1914),[1] and having peculiar merits for the work. For an understanding of these experiments and an appreciation of the conditions that make for their accuracy, it is necessary that the construction of this room be explained at some length. The west wing of the Jefferson Physical Laboratory is in plan a large square in the center of which rises a tower, which, for the sake of steadiness and insulation

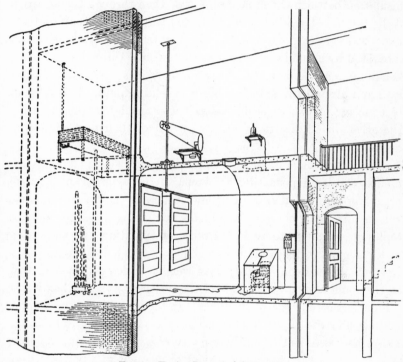

FIG. 2.   Testing Room and Apparatus

from all external vibration, is not merely of independent walls but has an entirely separate foundation, and above is spanned without touching by the roof of the main building. The sub-basement room of this tower is below the basement of the main building, but the walls of the latter are carried down to enclose it. The floor of the room is of concrete, the ceiling a masonry arch. There is but one door which leads through a small anteroom to the stairs mounting to the level of the basement of the main building. Through the

[1] See page 199, chapter 8.

ceiling there are two small openings for which special means of closing are provided. The larger of these openings barely permits the passage of an observer when raised or lowered by a block and tackle. It is necessary that there be some such entrance in order that observations may be taken in the room when the door is closed by the wall construction undergoing test.

Of prime importance, critical to the whole investigation, was the insulation between the rooms, otherwise than through the partition to be tested. The latter closed the doorway. Other than that the two rooms were separated by two eighteen-inch walls of brick, separated by a one-inch air space, not touching through a five-story height and carried down to separate foundations. Around the outer wall and around the antechamber was solid ground. It is difficult to conceive of two adjacent rooms better insulated, the one from the other, in all directions, except in that of their immediate connection.

The arrangement of apparatus, changed somewhat in later experiments, consisted primarily, as shown in the diagram, of a set of organ pipes, winded from a bellows reservoir in the room above, this in turn being charged from an air pump in a remote part of the building, — remote to avoid the noise of operation. In the center of the room two reflectors revolved slowly and noiselessly on roller bearings, turned continuously by a weight, under governor control, in the room above. The chair of the observer was in a box whose folding lids fitted over his shoulders. In the box was the small organ console and the key of the chronograph. The organ and chronograph had also console and key connection with the antechamber. The details of the apparatus are not of moment in a paper written primarily for architects.

Broadly, the method of measuring the transmission of sound through the partitions consisted in producing in the larger room a sound whose intensity in terms of threshold audibility was known, and reducing this intensity at a determinable rate until the sound ceased to be audible on the other side of the partition. The intensity of the sound at this instant was numerically equal to the reciprocal of the coefficient of transmission. This process involved several considerations which should at least be mentioned.

The sound of known intensity was produced by organ pipes of known powers of emission, allowance being made for the volume of the room, and the absorbing power of the walls. The method was fully explained in earlier papers.[1] It is to be borne in mind that there was thus determined merely the average of intensity. The intensity varied greatly in different parts of the room because of interference. In order that the average intensity of sound against the partition in a series of observations should equal the average intensity in the room, it was necessary to continuously shift the interference system. This was accomplished by means of revolving reflectors. This also rendered it possible to obtain a measure of average conditions in the room from observations taken in one position. Finally the observations in the room were always made by the observer seated in the box, as this rendered his clothing a negligible factor, and the condition of the room the same with or without his presence. Consideration was also given to the acoustical condition of the antechamber.

Two methods of reducing the sound have been employed. In the one the sound was allowed to die away naturally, the source being stopped suddenly, and the rate at which it decreased determined from the constants of the room. In another type of experiment the source, electrically maintained, was reduced by the addition of electrical resistance to the circuit. One method was suitable to one set of conditions, the other to another. The first was employed in the experiments whose results are given in this paper.

The first measurements were on felt, partly suggested by the experiments of Dr. Damrosch with felt on the walls of the Institute of Musical Art, partly because it offered the dynamically simplest problem on which to test the accuracy of the method by the concurrence of its results. The felt used was that so thoroughly studied in other acoustical aspects in the paper published in the Proceedings of the American Academy of Arts and Sciences in 1906. The door separating the two rooms was covered with a one-half inch thickness of this felt. The intensity of sound in the main room just audible through the felt was 3.7 times threshold audibility. Another layer of felt of equal thickness was added to the first, and the reduction in the

[1] See Reverberation, page 1.

intensity of sound in passing through the two was 7.8 fold. Through three-thickness, each one-half, the reduction was 15.4 fold, through four 30.4, five 47.5, and six 88.0. This test was for sounds having the pitch of violin C, first C above middle C, 512 vibrations per second.

There is another way of stating the above results which is perhaps of more service to architects. The ordinary speaking intensity of

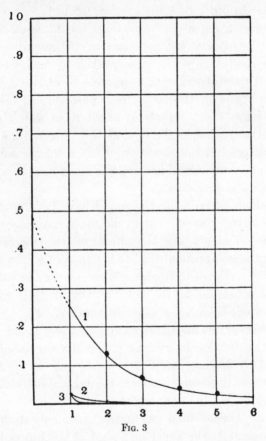

Fig. 3

the voice is — not exactly, of course, for it varies greatly — but of the order of magnitude of 1,000,000 times minimum audible intensity. Assume that there is a sound of that intensity, and of the pitch investigated, in a room in one side of a partition of half-inch felt. Its intensity on the other side of the partition would be 270,000 times minimum audible intensity. Through an inch of felt its intensity would be 128,000. Through six layers of such felt, that is, through three inches, its intensity would be 11,400 times minimum audible intensity, — very audible, indeed. The diminishing intensity of the sound as it proceeds through layer after layer of felt is plotted in the diagram (Curve 1, Fig. 3), in which all the points recorded are the direct results of observations. The intensity inside the room is the full ordinate of the diagram. The curve drawn is the nearest rectangular hyperbola fitting the observed and calculated points. The significance of this will be discussed later. It is sufficient for the present purpose to say that it is the theoretical curve for these conditions, and the close agreement between it and the observed points is a matter for considerable satisfaction.

The next partition tested was of sheet iron. This, of course, is not a normal building material and it may therefore seem disappointing and without interest to architects. But it is necessary to remember that these were preliminary investigations establishing methods and principles rather than practical data. Moreover, the material is not wholly impractical. The writer has used it in recommendations to an architect in one of the most interesting and successful cases of sound insulation so far undertaken — that in an after-theatre restaurant extending underneath the sidewalk of Broadway and 42d Street in New York.

The successive layers of sheet iron were held at a distance, each from the preceding, of one inch, spaced at the edges by a narrow strip of wood and felt, and pressed home by washers of felt. After the practical cases cited at the beginning of the paper, it requires courage and some hardihood to say that any insulation is good. It can only be said that every care was taken to this end. The results of the experiments can alone measure the efficiency of the method employed, and later they will be discussed with this in view.

The third series of experiments were with layers of sheet iron with one-half inch felt occupying part of the air space between them. The iron was that used in the second series, the felt that used in the first. The air space was unfortunately slightly greater than in the second series, being an inch and a quarter instead of an inch. The magnitude of the effect of this difference in distance was not realized at the time, but it was sufficient to prevent a direct com-

parison of the second and third series, and an attempt to deduce
the latter from the former with the aid of the first. When this was
realized, other conditions were so different as to make a repetition
of the series difficult.

In the following table is given the results of these three series of
experiments in such form as to admit of easy comparison. To this
end they are all reduced to the values which they would have had
with an intensity of sound in the inner room of 1,000,000. In the
first column each succeeding figure is the intensity outside an additional half inch of felt. In the second column, similarly, each succeeding figure is the intensity outside an additional sheet of iron.
In the third column, the second figure is the intensity outside a
single sheet of iron, and after that each succeeding figure is the
intensity outside of an additional felt and iron doublet with air space.

1,000,000	1,000,000	1,000,000
270,000	22,700	23,000
128,000	8,700	3,300
65,000	4,880	700
33,000	3,150	220
21,500	2,060	150
11,400	1,520	88

The sound transmitted in the second and third series is so much
less than in the first that when an attempt is made to plot it on the
same diagram (Curves 2 and 3, Fig. 3) it results in lines so low as to
be scarcely distinguishable from the base line. Magnifying the scale
tenfold (Fig. 4) throws the first series off the diagram for the earlier
values, but renders visible the second and third.

The method of representing the results of an investigation
graphically has several ends in view: it gives a visual impression of
the phenomenon; it shows by the nearness with which the plotted
values[1] lie to a smooth curve the accuracy of the method and of the
work; it serves to interpolate for intermediate values and to extrapolate for points which lie beyond the observed region, forward
or backward; finally, it reveals significant relations and leads to a

---

[1] In reproducing from the plotted diagrams for Figs. 3, 4, and 5, the dots, in some cases,
which indicated the plotted values of the observed points, do not clearly appear in distinction
on the lines. The greatest divergence, in any case, from the line drawn was not more than
twice the breadth of the line itself.

more effective discussion. It is worth while thus examining the three curves.

Attention has already been called to the curve for felt, to its extrapolation, and to the close approximation of the observed points to an hyperbola. The latter fact indicates the simplest possible law

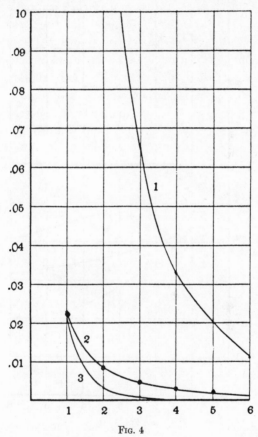

Fig. 4

of absorption. It proves that all layers absorb the same proportion of the sound; that each succeeding layer absorbs less actual sound than the preceding, but less merely because less sound reaches it to be absorbed. In the case in hand the sound in passing through the felt was reduced in the ratio 1.88 in each layer, 3.53 in each inch. It is customary to test such curves by plotting them on a special kind of coördinate paper, — one on which, while horizontal distances are uniformly scaled as before, vertical distances are scaled with greater and greater reduction, tenfold for each unit rise. On such coördinate paper the vertical distances are the power to which 10 must be raised to equal the number plotted — in other words, it is the logarithm of the number. Plotted on such paper the curve for

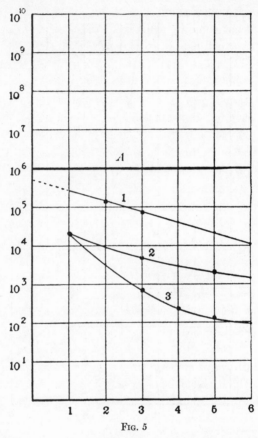

FIG. 5

felt will result in a straight line, if the curve in the other diagram was an hyperbola, and if the law of absorption was as inferred. How accurately it does so is shown in Curve 1, Fig. 5.

When the observations for iron, and for felt and iron, are similarly plotted (Curves 2 and 3, Fig. 5), the lines are not straight, but strongly curved upward, indicating that the corresponding curves in the preceding diagram were not hyperbolas, and that the law of constant coefficient did not hold. This must be explained in one or the other of two ways. Either there was some by-pass for the sound, or the efficiency of each succeeding unit of construction was less.

The by-pass as a possible explanation can be quickly disposed of. Take, for example, the extreme case, that for felt and iron, and make the extreme assumption that with the completed series of six screens all the sound has come by some by-pass, the surrounding walls, the foundations, the ceiling, or by some solid connection from the innermost to the outermost sheet. A calculation based on these assumptions gives a plot whose curvature is entirely at the lower end and bears no relationship to the observed values. In the other case, that of the iron only, a similar calculation gives a similar result; moreover, the much lower limit to which the felt and iron screens reduced the sound wholly eliminates any by-pass action as a vital factor in the iron-only experiment.

The other explanation is not merely necessary by elimination, but is dynamically rational. The screen walls such as here tested, as well as the screen walls in the actual construction described by way of introduction, do not act by absorption, as in the case of the felt; do not act by a process which is complete at the point, but rather by a process which in the first screen may be likened to reflection, and in the second and subsequent screens by a process which may be more or less likened to reflection, but which being in a confined space reacts on the screen or screens which have preceded it. In fact, the process must be regarded not as a sequence of independent steps or a progress of an independent action, but as that of a structure which must be considered dynamically as a whole.

When the phenomenon is one of pure absorption, as in felt, it is possible to express by a simple formula the intensity of the sound I, at any distance x, in terms of the initial intensity I_o,

$$I = I_o Rk^{-x},$$

where R represents the factor of surface discontinuity, and k the ratio in which the intensity is reduced in a unit distance. In the case of the felt tested, R is .485 and k is 3.53, the distance into the felt being measured in inches. As an application of this formula, one may compute the thickness of felt which would entirely extinguish a sound of the intensity of ordinary speech, — 10.4 inches. It is not possible to express by such a formula the transmission of sound through either of the more complex structures. However, it is possible to extrapolate empirically and show that 10.4 inches of neither would accomplish this ideal result, although they are both far superior to felt for thicknesses up to three inches in one case and five and one-half inches in the other.

A number of other experiments were tried during this preliminary stage of the investigation, such, for example, as increasing the distance between the screen walls, but it is not necessary to recount them here. Enough has already been given to show that a method had been developed for accurately measuring the insulating value of structures; more would but confuse the purpose. At this point the apparatus was improved, the method recast, and the investigation begun anew, thenceforward to deal only with standard forms of construction, and for sounds, not of one pitch only, but for the whole range of the musical scale.

11

WHISPERING GALLERIES

It is probable that all existing whispering galleries, it is certain that the six more famous ones, are accidents; it is equally certain that all could have been predetermined without difficulty, and like most accidents could have been improved upon. That these six, the Dome of St. Paul's Cathedral in London, Statuary Hall in the Capitol at Washington, the vases in the Salle des Cariatides in the Louvre in Paris, St. John Lateran in Rome, The Ear of Dionysius at Syracuse, and the Cathedral of Girgenti, are famous above others is in a measure due to some incident of place or association. Four are famous because on the great routes of tourist travel, one because of classical traditions, and one, in an exceedingly inaccessible city and itself still more inaccessible, through a curious story perpetuated by Sir John W. Herschel in the *Encyclopedia Metropolitana*. However, all show the phenomenon in a striking manner and merit the interest which they excite, an interest probably enhanced by the mystery attaching to an unpremeditated event in the five more modern cases, and none the less enhanced in the other by the tradition of its intentional design and as evidence of a "lost art."

The whispering gallery in the Capitol at Washington is of the simplest possible type.

The Capitol as first built was but the central portion of the present building, the Senate Chamber and the Hall of the House of Representatives being at that time immediately adjacent to the rotunda. With the admission of new states, and with the general increase in population, the Senate and the House outgrew their quarters, and in 1851 the great wings which now complete the building were constructed for their accommodation. The old Hall of the House, which in its day must have been acoustically an exceedingly poor assembly room, was transformed into the present Hall of Statues and became, or rather remained, one of the most perfect of whispering galleries.

The ceiling of the Hall of Statues, with the exception of a small circular skylight, is a portion of an exact sphere with its center very

Fig. 1. Hall of Statues, U. S. Capitol, Washington, D. C.

nearly at head level. As shown in the illustrations the ceiling is coffered. As originally constructed, and as it remained until 1901, the ceiling was perfectly smooth, being of wood, papered and painted in a manner to represent coffering. In 1901, a fire in the Chamber of the Supreme Court, also in the Capitol, led to a general overhauling of the building, and among other dangerous constructions the ceiling of wood in the Hall of Statues was replaced by a fireproof construction of steel and plaster. Instead of being merely painted, the new ceiling had recessed panels with mouldings and ribs in relief (Fig. 1). In consequence of this construction, the whispering gallery lost a large part of its unique quality.

During the years preceding the remodeling of the ceiling, the whispering gallery had been of great interest to tourists and deep hollows were worn in the marble tile where the observers stood. The experiment was usually tried in either one of two ways. The visitor to the gallery was placed at the center of curvature of the ceiling and told to whisper, when the slightest sounds were returned to him from the ceiling. The effect was much more striking than one would suppose from this simple description. The slight lapse of time required for the sound to travel to the ceiling and back, together with one's keen sense of direction, gave the effect of an invisible and mocking presence. Or the guide would place the tourists at symmetrical points on either side of the center, when they could with the help of the ceiling whisper to each other across distances over which they could not be heard directly. The explanation of this particular whispering gallery is exceedingly simple.

Speech, whether whispered or full toned, consists of waves or trains of waves of greatly varied character. The study, to its last refinement, of whispering gallery phenomena involves a consideration of this complicated character of speech, but a rough study, and one which serves most purposes, can be made by following the path and the transformation of a single wave. This can be illustrated by two series of photographs. In the one (Fig. 2), the wave is shown in the different stages of its advance outward, — spherical, except where it strikes the floor, the wall, or the repressed transverse arch of the ceiling. In the second series of photographs (Fig. 3), the wave has struck the spherical ceiling everywhere at the same instant,

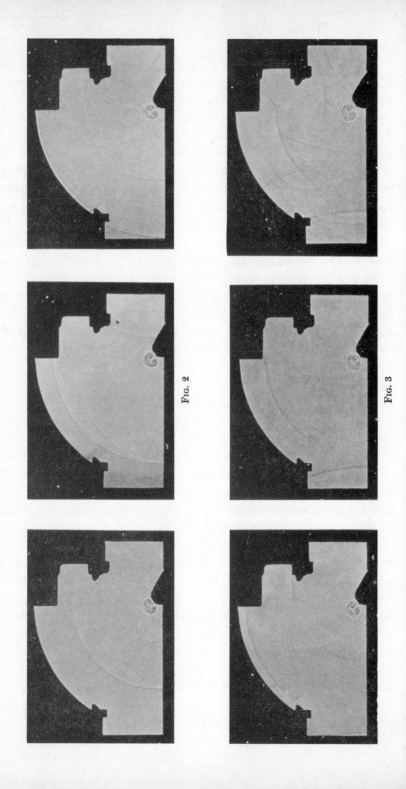

Fig. 2

Fig. 3

and, reversed in direction, gains in intensity as it gathers together toward the point from which it issued. The sound reflected from the other surfaces may be seen dividing and subdividing in multiple reflection and losing in intensity, while the sound reflected from the spherical ceiling gains through its rapid convergence.

These and other similar photographs used in this investigation were taken in a small sectional model, one-sixteenth of an inch to the foot in scale, made of plaster of Paris or of other convenient material, and the impulsive report or wave was produced either by the explosion of fulminate of mercury or directly by an electric spark. The flash by which the exposure was taken had a duration of less than a millionth of a second. It is wholly unnecessary for the purposes of this present discussion to go into the details of this process. It is sufficient to state that the illustrations are actual photographs of real sound-waves in the air and reproduce not merely the main but the subordinate phenomena.

In citing this gallery in an article on Whispering Galleries in Sturgis' *Dictionary of Architecture*, the writer made the statement that "The ceiling, painted so that it appears deeply panelled, is smooth. Had the ceiling been panelled the reflection would have been irregular and the effect very much reduced." A year or so after this was written the fire in the Capitol occurred, and in order to preserve the whispering gallery, which had become an object of unfailing interest to visitors to the Capitol, the new ceiling was made "to conform within a fraction of an inch" to the dimensions of the ceiling which it replaced. Notwithstanding this care, the quality of the room which had long made it the best and the best known of whispering galleries was in large measure lost. Since then this occurrence has been frequently cited as another of the mysteries of architectural acoustics and a disproof of the possibilities of predicting such phenomena. As a matter of fact, it was exactly the reverse. Only the part between the panels was reproduced in the original dimensions of the dome. The ceiling was no longer smooth, the staff was panelled in real recess and relief, and the result but confirmed the statement recorded nearly two years before in the *Dictionary of Architecture*.

The loss of this fine whispering gallery has at least some compensation in giving a convincing illustration, not merely of the conditions which make towards excellence in the phenomenon, but also of the conditions which destroy it. The effect of the paneling is obvious. Each facet on the complex ceiling is the source of a wavelet and as these facets are of different depths the resulting wavelets do not conspire to form the single focusing wave that results from a perfectly smooth dome. In a measure of course in this particular case the wavelets do conspire, for the reflecting surfaces are systematically placed and at one or the other of two or three depths. The dispersion of the sound, and the destruction of the whispering gallery is, therefore, not complete.

An instructive parallel may be drawn between acoustical and optical mirrors:

Almost any wall-surface is a much more perfect reflector of sound than the most perfect silver mirror is to light. In the former case, the reflection is over 96 per cent, in the latter case rarely over 90.

On the surfaces of the two mirrors scratches to produce equally injurious effects must be comparable in their dimensions to the lengths of the waves reflected. Audible sounds have wave lengths of from half an inch to sixty feet; visible light of from one forty-thousandth to one eighty-thousandth of an inch. Therefore while an optical mirror can be scratched to the complete diffusion of the reflected light by irregularities of microscopical dimensions, an acoustical mirror to be correspondingly scratched must be broken by irregularities of the dimensions of deep coffers, of panels, of engaged columns or of pilasters.

Moreover, just as remarkable optical phenomena are produced when the scratches on a mirror are parallel, equal, equal spaced, or of equal depth, as in mother of pearl, certain bird feathers, and in the optical grating, so also are remarkable acoustical phenomena produced when, as is usually the case in architectural construction, the relief and recess are equal, equally spaced, or of equal depth. The panels in the dome of the Hall of Statues of course diminish toward the apex of the dome and are thus neither equal nor equally spaced, but horizontally they are and produce corresponding phenomena. The full details of these effects are a matter of common knowledge in Physics but are not within the scope of the present

discussion. It is sufficient to say that the general result is a dispersion or a distortion in the form of the focus and that the general
effect is to greatly reduce the efficiency of the whispering gallery,
but to by no means wholly destroy it, as would be the case with
complete irregularity.

By the term whispering gallery is usually understood a room,
either artificial or natural, so shaped that faint sounds can be heard
across extraordinary distances. For this the Hall of Statues was ill-
adapted, partly because of a number of minor circumstances, but
primarily because a spherical surface is accurately adapted only to
return the sound directly upon itself. When the two points between
which the whisper is to be conveyed are separated, the correct form
of reflecting surface is an ellipsoid having the two points as foci.
When the two points are near together, the ellipsoid resembles more
and more a sphere, and the latter may be regarded as the limiting
case when the two points coincide. On the other hand, when the
two foci are very far apart the available part of the ellipsoid near one
of the foci resembles more and more a paraboloid, and this may be
regarded as the other extreme limiting case when one of the foci is at
an infinite or very great distance. I know of no building a considerable portion of whose wall or ceiling surface is part of an exact ellipsoid of revolution, but the great Mormon Tabernacle in Salt Lake
City is a near approximation. Plans of this remarkable building do
not exist, for it was laid out on the ground without the aid of formal
drawings soon after the settlers had completed their weary pilgrim-
age across the Utah desert and settled in their isolated valley. It was
built without nails, which were not to be had, and held together
merely by wooden pins and tied with strips of buffalo hide. Notwithstanding this construction, and notwithstanding the fact that it
spans 250 feet in length, and 150 feet in breadth, and is without any
interior columns of any sort, it has been free from the necessity of
essential repair for over fifty years. As the photograph (Fig. 5)
shows, taken at the time of building, the space between the ceiling
and the roof is a wooden bridge truss construction. These photographs, given by the elders of the church, are themselves interesting considering the circumstances under which they were taken,
the early date and the remote location.

FIG. 4. Exterior, Mormon Tabernacle, Salt Lake City, Utah.

FIG. 5. Photograph showing Construction, Mormon Tabernacle, Salt Lake City, Utah.

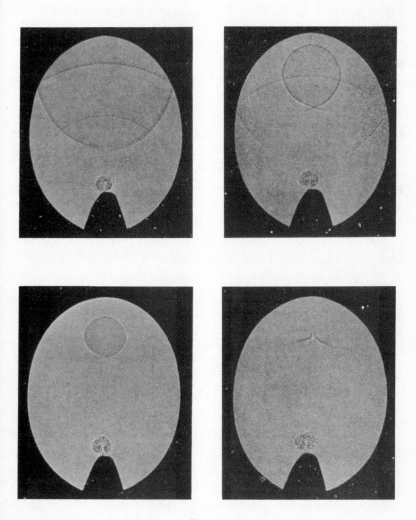

F<small>IG</small>. 6

It is difficult for an interior photograph of a smooth ceiling to give an impression of its shape. An idea of the shape of the interior of the Tabernacle may be obtained, however, from a photograph of its exterior. It obviously somewhat resembles an ellipsoid of revolution. It is equally obvious that it is not exactly that. Nevertheless there are two points between which faint sounds are carried with remarkable distinctness, — the reader's desk and the front of the balcony in the rear.

The essential geometrical property of an ellipsoid of revolution is that lines drawn to any point of the surface from the two foci make equal angles with the surface. It follows that sound diverging from one focus will be reflected toward the other. The preceding photographs (Fig. 6) show the progress of a sound-wave in the model of an idealized whispering gallery of this type in which the reflecting surface is a portion of a true ellipsoid of revolution.

The most notable whispering gallery of this type is that described by Sir John Herschel in one of the early scientific encyclopedias, the *Encyclopedia Metropolitana* as follows:

In the Cathedral of Girgenti in Sicily, the slightest whisper is borne with perfect distinctness from the great western floor to the cornice behind the high altar, a distance of 250 feet. By a most unlucky coincidence the precise focus of divergence at the former station was chosen for the place of the confessional. Secrets never intended for the public ear thus became known, to the dismay of the confessor and the scandal of the people, by the resort of the curious to the opposite point, which seems to have been discovered by accident. . . .

Aside from the great distance between the foci, the circumstances related had many elements of improbability and the final discussion of this subject was postponed from year to year in the hope that the summer's work, which has usually been devoted to the study of European auditoriums, would carry the writer near Girgenti, an interesting but rather inaccessible city on the southwestern coast of Sicily. Finally, failing any especially favorable opportunity, a flying trip was made from the north of Europe with the study of this gallery and of the Ear of Dionysius at Syracuse as the sole objective. On the way down the perplexity of the case was increased by finding in Baedeker the statement that there is a noteworthy whispering gal-

Fig. 7.   Interior, Cathedral of Girgenti, Sicily.

lery between the west entrance of the Cathedral and "the steps of the high altar." Such a whispering gallery is wholly inconceivable. The facts showed a whispering gallery between the foci as described by Herschel, although the accompanying story is rendered improbable by the extreme inaccessibility of the more remote focus, and its very conspicuous position. Nor is the distance so great as stated by Herschel, being a little over 100 feet instead of 250 feet. However, the interest in this whispering gallery arises not because of any incident attending its discovery, but because it illustrates, albeit rather crudely, the form of surface giving the best results for whispering between two very widely separated points.

As already stated the strictly correct form of surface for a whispering gallery is an ellipsoid of revolution whose foci coincide with the two points between which there is to be communication. In the whispering gallery in the Cathedral of Girgenti (Fig. 7), the focusing surface consists of a quarter of a sphere prolonged in the shape of a half cylinder forming the ceiling over the chancel. This is obviously not a true paraboloid, and, such as it is, it is interrupted by an arch of slight reveal where the cylinder joins the sphere; moreover, the two points of observation do not lie on the axis of revolution as they should for the best result. But a hemisphere and a continuing cylinder make a fair approach to a portion of a paraboloid; and while the two points of observation are not on the axis of revolution, they are on a secondary axis, the station by the door being below, and the focus in the chancel being at a corresponding distance above the principal axis.

In all the preceding galleries, there is but a single reflection between the radiant and the receiving foci. There are others in which there are several such reflections. Well-known examples are the church of St. John Lateran in Rome and in the Salle des Cariatides in the Louvre.

In the Church of St. John Lateran (Fig. 8), each bay in the great side aisles is a square having a ceiling which is approximately a portion of a sphere. At best, the approximation of the ceiling to a sphere is not close and the ceiling varies from bay to bay, not intentionally but merely as a matter of variation in construction. In one bay more closely than in the others the ceiling, regarded as an acoustical

FIG. 8.　Interior, Church of St. John Lateran, Rome.

mirror, has its foci nearly at head level. In consequence of this, two observers standing at opposite corners can whisper to each other with the ceiling as a reflecting surface. The curvature even in this bay is not ideal for the production of a whispering gallery, so that thus used the gallery is far from notable. It so happens, however, that the great square columns which form the corners of each bay have, instead of sharp corners, a reëntering cove or fluting in the arc of a circle and over twelve inches across in opening. If the observers, instead of attempting to speak directly to the ceiling, turn back to back and face the columns standing close to them, this great fluting gathers the sound from the speaker and directs it in a concentrated cone to the ceiling; this returning from the ceiling to the opposite angle of the bay is concentrated by the opposite fluting on the other observer. In more scientific language, borrowed from the nomenclature of the makers of optical instruments, the flutings increase the angular aperture of the system.

An almost exact duplicate of this whispering gallery is to be found in the vestibule of the Conservatoire des Arts et Metiers in Paris. This vestibule, itself also an exhibition room but called since the discovery of its peculiar property La Salle-Echo, is square with rounded corners and a low domical ceiling. Here, as in St. John Lateran, the observers face the corners and the whisper undergoes three reflections between the foci. The fact that the two observers are back to back diminishes the sound which would otherwise pass directly between them and makes the whispering gallery more pronounced and the phenomenon much more striking. In both galleries it is the custom for the observers to take their positions in a somewhat random manner. The correct position is at a distance from the concave cylindrical surface a little less than half the radius of curvature.

In these whispering galleries the surfaces are not theoretically correct and the phenomenon is far from perfect. This failure of loudness and distinctness in most of the multiple reflection galleries arises not from any progressive loss in the many reflections, for the loss of energy in reflection is practically negligible. Indeed, given ideally shaped surfaces, multiple reflection whispering galleries are capable of producing exceptional effect; for if two of the surfaces be very near the observers they may, even though they themselves be of

FIG. 9.   Salle des Cariatides, the Louvre, Paris.

small dimensions, gather into the phenomenon very large portions of the emergent and of the focused whisper. In both St. John Lateran and La Salle-Echo, the condensing mirrors are cylindrical and gather the sound horizontally only. In the vertical plane, they are wholly without effect.

It is not difficult to determine the correct forms for the extreme mirrors. If the ceiling be flat, the reflecting surfaces near the two observers should be parabolic with the axis of the paraboloid directed toward the center of the ceiling, the correct position for the mouth of the speaker and the ear of the auditor being at the foci of the two paraboloids. If the ceiling be curved, the simplest design is when the first and last reflectors are portions of an ellipsoid, each with one focus at the center of the ceiling and the other at one of the foci of the system as a whole. Finally, if the ceiling be curved, there is still another theoretical shape for the end reflectors, determined by the curvature of the ceiling; in this case the ideal surface is not a conic surface, nor otherwise geometrically simple, but is such that the converging power of the end mirror with half the converging power of the middle mirror will give a plane wave.

It is obvious that the accurate fulfilling of these conditions by accident is improbable, but they are at least approached in the whispering gallery in the Salle des Cariatides in the Louvre (Fig. 9). Along the axis of the room, and at no inconsiderable distance apart, are two large shallow antique vases. A whisper uttered a little within the rim of one is partially focused by it, is still further focused by the barrel-shaped ceiling, and is brought to a final focus symmetrically within the rim of the further vase. It is evident that the effect is dependent on only a portion of each vase, but this portion satisfies the necessary conditions to a first approximation in both longitudinal and in transverse section. When the correct foci are found this whispering gallery is very distinct in its enunciation. It would be even more distinct if the ceiling of the room were slightly lower, or, keeping the height the same, if its radius of curvature were slightly greater. It would be still better if the vases were slightly deeper.

The whispering gallery which has received the greatest amount of discussion, and a discussion curiously inadequate in view of the eminence of the authorities engaged, is the circular gallery at the base of

Fig. 10.   Section through Dome of St. Paul's Cathedral, London.

the dome of St. Paul's Cathedral in London. This gallery was first brought into scientific consideration by Sir John Herschel, who in describing it stated that "the faintest sound is faithfully conveyed from one side to the other of the dome, but is not heard at any intermediate point." According to Lord Rayleigh, whose reference, however, I am unable to verify, and either in page or edition must be in error, an early explanation of this was by Sir George Airy, the Astronomer Royal, who "ascribed it to the reflection from the surface of the dome overhead." Airy could have been led into such error only by the optical illusion whereby a dome seen from within seems lower than it is in reality. A moment's inspection of the preceding illustration (Fig. 10), which the Clerk of the Works kindly had reproduced from an old engraving in the possession of the cathedral, shows that this explanation would be incorrect. The guide who does the whispering usually occupies the position marked "A"; the other focus is in the position marked "B." The focus accounted for by Airy would be high up in the dome. Lord Rayleigh taking exception both to the statement of fact by Herschel and the explanation by Airy wrote "I am disposed to think that the principal phenomenon is to be explained somewhat differently. The abnormal loudness with which a whisper is heard is not confined to the position diametrically opposite to that occupied by the whisperer, and therefore, it would appear, does not depend materially upon the symmetry of the dome. The whisper seems to creep around the gallery horizontally, not necessarily along the shorter arc, but rather along that arc toward which the whisperer faces. This is in consequence of the very unequal audibility of a whisper in front of and behind the speaker, a phenomenon which may easily be observed in the open air." Lord Rayleigh's explanation of the phenomenon in this case as due to the "creeping" of the sound around the circular wall immediately surrounding the narrow gallery accessible to visitors is unquestionably correct. It is but another way of phrasing this explanation to say that the intensification of the sound is due to its accumulation when turned on itself by the restraining wall. It is obvious that the main intensification arises from the curved wall returning on itself. Vertically, the sound spreads almost as it would were the curved wall developed on a plane. This vertical spreading of the sound is in a measure restricted by the circular floor gallery and by the overhanging ledge of the cornice moulding. The cornice can be made to contribute most to the effect by making the curve of its lines below the principal projecting ledge, that which corresponds to the drip moulding of an exterior cornice, relatively smooth and simple.

But even Lord Rayleigh's explanation does not fully account for the truly remarkable qualities of this whispering gallery. There are many circular walls as high, as hard, and as smooth as that in St. Paul's Gallery but in which the whispering gallery is not to be compared in quality. The rear walls of many semi-circular auditoriums satisfy these conditions without producing parallel results, for example in the Fogg lecture-room at Harvard University before it was altered, and in the auditorium just completed at Cornell University. A feature of the whispering gallery in St. Paul's, contributing not a little to its efficiency, is the inclination of its wall, less noticeable perhaps in the actual gallery than in the architectural " Section." The result is that all the sound which passes the quarter point of the gallery, the point half way around between the foci, is brought down to the level of the observer, and, combined with the reflection from the ledge which constitutes the broad seat running entirely around the gallery, confines and intensifies the sound. This feature is of course of unusual occurrence.

It may not be out of place to give the dimensions of this gallery. The distance from focus to focus, if indeed in this type of gallery they can be called foci, is 150 feet. The wall has a height of 20 feet, and is not moulded in panels as shown in the engraving, but is smooth except for eight shallow niches. While the inclination of the wall in the gallery of St. Paul's is a contributing factor, an even more efficient wall would have been one very slightly, indeed almost imperceptibly, curved, the section being the arc of a circle struck from the center of the dome on a level with the observers. Such a gallery will be in the dome of the Missouri State Capitol, a gallery unique in this respect that it will have been planned intentionally by the architects.[1]

A discussion of noted whispering galleries would not be complete

[1] The building is now complete One of the architects, Mr. Edgerton Swartwout, reports that the whispering gallery in the dome exactly fulfills Professor Sabine's prediction, and has been the cause of much curiosity and astonishment. — Editor.

without mention of the famous Ear of Dionysius at Syracuse. A mile out from the present city of Syracuse, on the slope of the terrace occupied by the Neapolis of the ancient city, are the remains of a quarry entered on one side on the level but cut back to perpendicular walls from a hundred to a hundred and thirty feet in height. This old quarry, now overgrown by a wild and luxurious vegetation, is known as the Latomia del Paradiso. At its western angle is a great grotto, shaped somewhat like an open letter S, 210 feet in winding length, 74 feet high, 35 feet in width at the base and narrowing rapidly toward the top. The innermost end of this grotto is nearly circular, and the rear wall slopes forward as it rises preserving in revolution the same contour that characterizes the two sides throughout their length. The top is a narrow channel of a uniform height and but a few feet in width. At the innermost end of this channel, at the apex of the half cone which forms the inner end of the grotto, is a vertical opening four or five feet square, scarcely visible, certainly not noticeable, from below. This opening is into a short passageway

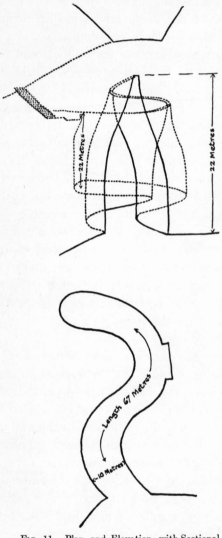

Fig. 11. Plan and Elevation, with Sectional Indication, of Ear of Dionysius, Syracuse, Sicily.

Fig. 12. View of Outer Opening, the So-called Ear of Dionysius, Syracuse, Sicily.

which leads to a flight of steps and thence to the ground above (Fig.
11). The grotto is noted for two somewhat inconsistent acoustical
properties. When being shown the grotto from below, one's attention is called to its very remarkable reverberation. When above,
one's attention is called to the ability to hear what is said at any
point on the floor.

It is related that Tyrant Dionysius, the great builder of Syracuse,
so designed his prisons that at certain concealed points of observation
he could not merely see everything that was done, but, through remarkable acoustical design, could hear every word which was spoken,
even when whispered only (Fig. 12). There is a tradition, dating
back however only to the sixteenth century, that this grotto, since
then called the Ear of Dionysius, was such a prison. Quarries were
plausible prisons in which captives of war might have been com-
pelled to work, and there are, surrounding this quarry, traces of a
wall and sentry houses, but there is no direct evidence associating
this grotto with Dionysius, unless indeed one regards its interesting
acoustical properties taken in connection with classical tradition as
such evidence.

In its acoustical property this grotto resembles more a great ear
trumpet than a whispering gallery in the ordinary sense of the word.
It is, of course, in no sense a focusing whispering gallery of the type
represented by the vases and curved ceiling in the Louvre. It more
nearly resembles the gallery in St. Paul's Cathedral, but the sound
is not spoken close to the deflecting wall, one of the essentially
characteristic conditions of a true whispering gallery of that type,
and the wall is not continuously concave. In fact, in other ways also
its acoustical property is not very notable, for distinctness of enunciation is blurred by excessive reverberation.

It is conceivable that whispering galleries should be of use and
purposeful, but it is more probable that they will remain architectural
curiosities. When desired, they may be readily woven into the design
of many types of monumental buildings.

APPENDIX

Dᴜʀɪɴɢ one of the early lectures given at the Sorbonne in the spring
of 1917, Professor Sabine referred to the difficulties inherent in experiments on sound intensities. The following is a free translation
from the notes, in French, which he prepared for this lecture:

In no other doman have physicists disregarded the conditions introduced by the surrounding materials, but in acoustics these do not
seem to have received the least attention. If measurements are made
in the open air, over a lawn, as was done by Lord Rayleigh in certain
experiments, is due consideration given to the fact that the surface
has an absorbing power for sound of from 40 to 60 per cent? Or, if inside a building, as in Wien's similar experiments, is allowance made
for the fact that the walls reflect from 93 to 98 per cent of the sound?
We need not be surprised if the results of such experiments differ
from one another by a factor of more than a hundred.

It would be no more absurd to carry out photometric measurements in a room where the walls, ceiling, and even the floor and tables
consisted of highly polished mirrors, than to make measurements on
the intensity, or on the quantitative analysis of sound, under the conditions in which such experiments have almost invariably been executed. It is not astonishing that we have been discouraged by the
results, and that we may have despaired of seeing acoustics occupy
the position to which it rightly belongs among the exact sciences.

The length of the waves of light is so small compared with the
dimensions of a photometer that we do not need to concern ourselves
with the phenomena of interference while measuring the intensity of
light. In the case of sound, however, it must be quite a different
matter.

In order to show this in a definite manner, I have measured the
intensity of the sound in all parts of a certain laboratory room. For
simplicity, a symmetrical room was selected, and the source, giving a
very pure tone, was placed in the center. It was found that, near the

source, even at the source itself, the intensity was in reality less than at a distance of five feet from the source. And yet, the clever experimenter, Wien, and the no less skillful psychologists Wundt and Münsterberg have assumed under similar conditions the law of variation of intensity with the inverse square of the distance. It makes one wonder how they were able to draw any conclusions from their measurements.

Not only do the walls reflect sound in such a way that it becomes many times more intense than it otherwise would be; and not only does the interference of sound exist to such an extent that we find regions of maximum and regions of minimum of sound in a room; but even the total quantity of sound emitted by the source itself may be greatly affected by its position with regard to the interference system of the room.

This will be more readily understood if illustrated by an incident drawn from the actual experiments. A special sort of felt, of strong absorbing power, was brought into the room and placed on the floor. The effect was two-fold. First, the introduction of the felt increased the absorption of the sound, and thus tended to diminish the total intensity of sound in the room, theoretically to a third of its previous value. But actually it had the contrary effect; the sound became much louder than before. The felt was so placed on the floor as to shift the interference system in the room, and thus the reaction of the sound vibrations in the room upon the source itself was modified. The source was a vibrating diaphragm situated at the base of a resonating chamber. In its first location, the source was at a node of condensation, where the motion of the sound which had accumulated in the room coincided with that of the diaphragm. It was thus difficult for the diaphragm to impart any additional motion to the air. In the second case, however, the vibrations of the two were opposite; the diaphragm was able to push upon the air, and although the amplitude of its motion was somewhat reduced by the reaction of the air upon it, the emitted sound was louder. When under these conditions the diaphragm was forced to vibrate with the same amplitude as at first, the emitted sound became eight times louder.

Naturally these two positions in the interference system were designedly selected, and they show exceptional reactions on the source.

However, in the case of a very complex sound, a comparable divergence in the reaction of the room on the different components of the sound would be probable.

It is thus necessary in quantitative research in acoustics to take account of three factors: the effect of reflection by the walls on the increase of the total intensity of sound in the room; the effect of interference in greatly altering the distribution of this intensity; and the effect of the reaction of the sound vibrations in a room upon the source itself. . . .

In choosing a source of sound, it has usually been assumed that a source of fixed amplitude was also a source of fixed intensity, e. g., a vibrating diaphragm or a tuning fork electrically maintained. On the contrary, this is just the sort of source whose emitting power varies with the position in which it is placed in the room. On the other hand, an organ pipe is able within certain limits to adjust itself automatically to the reaction due to the interference system. We may say, simply, that the best standard source of sound is one in which the greatest percentage of emitted energy takes the form of sound.

CATALOG OF DOVER BOOKS

BOOKS EXPLAINING SCIENCE AND MATHEMATICS

THE COMMON SENSE OF THE EXACT SCIENCES, W. K. Clifford. Introduction by James Newman, edited by Karl Pearson. For 70 years this has been a guide to classical scientific and mathematical thought. Explains with unusual clarity basic concepts, such as extension of meaning of symbols, characteristics of surface boundaries, properties of plane figures, vectors, Cartesian method of determining position, etc. Long preface by Bertrand Russell. Bibliography of Clifford. Corrected, 130 diagrams redrawn. 249pp. 5⅜ x 8.
T61 Paperbound $1.60

SCIENCE THEORY AND MAN, Erwin Schrödinger. This is a complete and unabridged reissue of SCIENCE AND THE HUMAN TEMPERAMENT plus an additional essay: "What is an Elementary Particle?" Nobel Laureate Schrödinger discusses such topics as nature of scientific method, the nature of science, chance and determinism, science and society, conceptual models for physical entities, elementary particles and wave mechanics. Presentation is popular and may be followed by most people with little or no scientific training. "Fine practical preparation for a time when laws of nature, human institutions . . . are undergoing a critical examination without parallel," Waldemar Kaempffert, N. Y. TIMES. 192pp. 5⅜ x 8.
T428 Paperbound $1.35

PIONEERS OF SCIENCE, O. Lodge. Eminent scientist-expositor's authoritative, yet elementary survey of great scientific theories. Concentrating on individuals—Copernicus, Brahe, Kepler, Galileo, Descartes, Newton, Laplace, Herschel, Lord Kelvin, and other scientists—the author presents their discoveries in historical order adding biographical material on each man and full, specific explanations of their achievements. The clear and complete treatment of the post-Newtonian astronomers is a feature seldom found in other books on the subject. Index. 120 illustrations. xv + 404pp. 5⅜ x 8.
T716 Paperbound $1.50

THE EVOLUTION OF SCIENTIFIC THOUGHT FROM NEWTON TO EINSTEIN, A. d'Abro. Einstein's special and general theories of relativity, with their historical implications, are analyzed in non-technical terms. Excellent accounts of the contributions of Newton, Riemann, Weyl, Planck, Eddington, Maxwell, Lorentz and others are treated in terms of space and time, equations of electromagnetics, finiteness of the universe, methodology of science. 21 diagrams. 482pp. 5⅜ x 8.
T2 Paperound $2.00

THE RISE OF THE NEW PHYSICS, A. d'Abro. A half-million word exposition, formerly titled THE DECLINE OF MECHANISM, for readers not versed in higher mathematics. The only thorough explanation, in everyday language, of the central core of modern mathematical physical theory, treating both classical and modern theoretical physics, and presenting in terms almost anyone can understand the equivalent of 5 years of study of mathematical physics. Scientifically impeccable coverage of mathematical-physical thought from the Newtonian system up through the electronic theories of Dirac and Heisenberg and Fermi's statistics. Combines both history and exposition; provides a broad yet unified and detailed view, with constant comparison of classical and modern views on phenomena and theories. "A must for anyone doing serious study in the physical sciences," JOURNAL OF THE FRANKLIN INSTITUTE. "Extraordinary faculty . . . to explain ideas and theories of theoretical physics in the language of daily life," ISIS. First part of set covers philosophy of science, drawing upon the practice of Newton, Maxwell, Poincaré, Einstein, others, discussing modes of thought, experiment, interpretations of causality, etc. In the second part, 100 pages explain grammar and vocabulary of mathematics, with discussions of functions, groups, series, Fourier series, etc. The remainder is devoted to concrete, detailed coverage of both classical and quantum physics, explaining such topics as analytic mechanics, Hamilton's principle, wave theory of light, electromagnetic waves, groups of transformations, thermodynamics, phase rule, Brownian movement, kinetics, special relativity, Planck's original quantum theory, Bohr's atom, Zeeman effect, Broglie's wave mechanics, Heisenberg's uncertainty, Eigen-values, matrices, scores of other important topics. Discoveries and theories are covered for such men as Alembert, Born, Cantor, Debye, Euler, Foucault, Galois, Gauss, Hadamard, Kelvin, Kepler, Laplace, Maxwell, Pauli, Rayleigh, Volterra, Weyl, Young, more than 180 others. Indexed. 97 illustrations. ix + 982pp. 5⅜ x 8.
T3 Volume 1, Paperbound $2.00
T4 Volume 2, Paperbound $2.00

CONCERNING THE NATURE OF THINGS, Sir William Bragg. Christmas lectures delivered at the Royal Society by Nobel laureate. Why a spinning ball travels in a curved track; how uranium is transmuted to lead, etc. Partial contents: atoms, gases, liquids, crystals, metals, etc. No scientific background needed; wonderful for intelligent child. 32pp. of photos, 57 figures. xii + 232pp. 5⅜ x 8.
T31 Paperbound $1.35

THE UNIVERSE OF LIGHT, Sir William Bragg. No scientific training needed to read Nobel Prize winner's expansion of his Royal Institute Christmas Lectures. Insight into nature of light, methods and philosophy of science. Explains lenses, reflection, color, resonance, polarization, x-rays, the spectrum, Newton's work with prisms, Huygens' with polarization, Crookes' with cathode ray, etc. Leads into clear statement of 2 major historical theories of light, corpuscle and wave. Dozens of experiments you can do. 199 illus., including 2 full-page color plates. 293pp. 5⅜ x 8.
S538 Paperbound $1.85

PHYSICS, THE PIONEER SCIENCE, L. W. Taylor. First thorough text to place all important physical phenomena in cultural-historical framework; remains best work of its kind. Exposition of physical laws, theories developed chronologically, with great historical, illustrative experiments diagrammed, described, worked out mathematically. Excellent physics text for self-study as well as class work. Vol. 1: Heat, Sound: motion, acceleration, gravitation, conservation of energy, heat engines, rotation, heat, mechanical energy, etc. 211 illus. 407pp. 5⅜ x 8. Vol. 2: Light, Electricity: images, lenses, prisms, magnetism, Ohm's law, dynamos, telegraph, quantum theory, decline of mechanical view of nature, etc. Bibliography. 13 table appendix. Index. 551 illus. 2 color plates. 508pp. 5⅜ x 8.

Vol. 1 S565 Paperbound **$2.00**
Vol. 2 S566 Paperbound **$2.00**
The set **$4.00**

FROM EUCLID TO EDDINGTON: A STUDY OF THE CONCEPTIONS OF THE EXTERNAL WORLD, Sir Edmund Whittaker. A foremost British scientist traces the development of theories of natural philosophy from the western rediscovery of Euclid to Eddington, Einstein, Dirac, etc. The inadequacy of classical physics is contrasted with present day attempts to understand the physical world through relativity, non-Euclidean geometry, space curvature, wave mechanics, etc. 5 major divisions of examination: Space; Time and Movement; the Concepts of Classical Physics; the Concepts of Quantum Mechanics; the Eddington Universe. 212pp. 5⅜ x 8.   T491 Paperbound **$1.35**

THE STORY OF ATOMIC THEORY AND ATOMIC ENERGY, J. G. Feinberg. Wider range of facts on physical theory, cultural implications, than any other similar source. Completely non-technical. Begins with first atomic theory, 600 B.C., goes through A-bomb, developments to 1959. Avogadro, Rutherford, Bohr, Einstein, radioactive decay, binding energy, radiation danger, future benefits of nuclear power, dozens of other topics, told in lively, related, informal manner. Particular stress on European atomic research. "Deserves special mention . . . authoritative," Saturday Review. Formerly "The Atom Story." New chapter to 1959. Index. 34 illustrations. 251pp. 5⅜ x 8.   T625 Paperbound **$1.45**

THE STRANGE STORY OF THE QUANTUM, AN ACCOUNT FOR THE GENERAL READER OF THE GROWTH OF IDEAS UNDERLYING OUR PRESENT ATOMIC KNOWLEDGE, B. Hoffmann. Presents lucidly and expertly, with barest amount of mathematics, the problems and theories which led to modern quantum physics. Dr. Hoffmann begins with the closing years of the 19th century, when certain trifling discrepancies were noticed, and with illuminating analogies and examples takes you through the brilliant concepts of Planck, Einstein, Pauli, de Broglie, Bohr, Schroedinger, Heisenberg, Dirac, Sommerfeld, Feynman, etc. This edition includes a new, long postscript carrying the story through 1958. "Of the books attempting an account of the history and contents of our modern atomic physics which have come to my attention, this is the best," H. Margenau, Yale University, in "American Journal of Physics." 32 tables and line illustrations. Index. 275pp. 5⅜ x 8.   T518 Paperbound **$1.45**

SPACE AND TIME, Emile Borel. An entirely non-technical introduction to relativity, by world-renowned mathematician, Sorbonne Professor. (Notes on basic mathematics are included separately.) This book has never been surpassed for insight, and extraordinary clarity of thought, as it presents scores of examples, analogies, arguments, illustrations, which explain such topics as: difficulties due to motion; gravitation a force of inertia; geodesic lines; wave-length and difference of phase; x-rays and crystal structure; the special theory of relativity; and much more. Indexes. 4 appendixes. 15 figures. xvi + 243pp. 5⅜ x 8.
T592 Paperbound **$1.45**

THE RESTLESS UNIVERSE, Max Born. New enlarged version of this remarkably readable account by a Nobel laureate. Moving from sub-atomic particles to universe, the author explains in very simple terms the latest theories of wave mechanics. Partial contents: air and its relatives, electrons & ions, waves & particles, electronic structure of the atom, nuclear physics. Nearly 1000 illustrations, including 7 animated sequences. 325pp. 6 x 9.
T412 Paperbound **$2.00**

SOAP SUBBLES, THEIR COLOURS AND THE FORCES WHICH MOULD THEM, C. V. Boys. Only complete edition, half again as much material as any other. Includes Boys' hints on performing his experiments, sources of supply. Dozens of lucid experiments show complexities of liquid films, surface tension, etc. Best treatment ever written. Introduction. 83 illustrations. Color plate. 202pp. 5⅜ x 8.   T542 Paperbound **95¢**

SPINNING TOPS AND GYROSCOPIC MOTION, John Perry. Well-known classic of science still unsurpassed for lucid, accurate, delightful exposition. How quasi-rigidity is induced in flexible and fluid bodies by rapid motions; why gyrostat falls, top rises; nature and effect on climatic conditions of earth's precessional movement; effect of internal fluidity on rotating bodies, etc. Appendixes describe practical uses to which gyroscopes have been put in ships, compasses, monorail transportation. 62 figures. 128pp. 5⅜ x 8. T416 Paperbound **$1.00**

MATTER & LIGHT, THE NEW PHYSICS, L. de Broglie. Non-technical papers by a Nobel laureate explain electromagnetic theory, relativity, matter, light and radiation, wave mechanics, quantum physics, philosophy of science. Einstein, Planck, Bohr, others explained so easily that no mathematical training is needed for all but 2 of the 21 chapters. Unabridged. Index. 300pp. 5⅜ x 8.   T35 Paperbound **$1.60**

BRIDGES AND THEIR BUILDERS, David Steinman and Sara Ruth Watson. Engineers, historians, everyone who has ever been fascinated by great spans will find this book an endless source of information and interest. Dr. Steinman, recipient of the Louis Levy medal, was one of the great bridge architects and engineers of all time, and his analysis of the great bridges of history is both authoritative and easily followed. Greek and Roman bridges, medieval bridges, Oriental bridges, modern works such as the Brooklyn Bridge and the Golden Gate Bridge, and many others are described in terms of history, constructional principles, artistry, and function. All in all this book is the most comprehensive and accurate semipopular history of bridges in print in English. New, greatly revised, enlarged edition. 23 photographs, 26 line drawings. Index. xvii + 401pp. 5⅜ x 8.   T431 Paperbound **$2.00**

FADS AND FALLACIES IN THE NAME OF SCIENCE, Martin Gardner. Examines various cults, quack systems, frauds, delusions which at various times have masqueraded as science. Accounts of hollow-earth fanatics like Symmes; Velikovsky and wandering planets; Hoerbiger; Bellamy and the theory of multiple moons; Charles Fort; dowsing, pseudoscientific methods for finding water, ores, oil. Sections on naturopathy, iridiagnosis, zone therapy, food fads, etc. Analytical accounts of Wilhelm Reich and orgone sex energy; L. Ron Hubbard and Dianetics; A. Korzybski and General Semantics; many others. Brought up to date to include Bridey Murphy, others. Not just a collection of anecdotes, but a fair, reasoned appraisal of eccentric theory. Formerly titled IN THE NAME OF SCIENCE. Preface. Index. x + 384pp. 5⅜ x 8.   T394 Paperbound **$1.50**

See also: **A PHILOSOPHICAL ESSAY ON PROBABILITIES, P. de Laplace; ON MATHEMATICS AND MATHEMATICIANS, R. E. Moritz; AN ELEMENTARY SURVEY OF CELESTIAL MECHANICS, Y. Ryabov; THE SKY AND ITS MYSTERIES, E. A. Beet; THE REALM OF THE NEBULAE, E. Hubble; OUT OF THE SKY, H. H. Nininger; SATELLITES AND SCIENTIFIC RESEARCH, D. King-Hele; HEREDITY AND YOUR LIFE, A. M. Winchester; INSECTS AND INSECT LIFE, S. W. Frost; PRINCIPLES OF STRATIGRAPHY, A. W. Grabau; TEACH YOURSELF SERIES.**

HISTORY OF SCIENCE AND MATHEMATICS

DIALOGUES CONCERNING TWO NEW SCIENCES, Galileo Galilei. This classic of experimental science, mechanics, engineering, is as enjoyable as it is important. A great historical document giving insights into one of the world's most original thinkers, it is based on 30 years' experimentation. It offers a lively exposition of dynamics, elasticity, sound, ballistics, strength of materials, the scientific method. "Superior to everything else of mine," Galileo. Trans. by H. Crew, A. Salvio. 126 diagrams. Index. xxi + 288pp. 5⅜ x 8.
S99 Paperbound **$1.65**

A DIDEROT PICTORIAL ENCYCLOPEDIA OF TRADES AND INDUSTRY, Manufacturing and the Technical Arts in Plates Selected from "L'Encyclopédie ou Dictionnaire Raisonné des Sciences, des Arts, et des Métiers" of Denis Diderot. Edited with text by C. Gillispie. This first modern selection of plates from the high point of 18th century French engraving is a storehouse of valuable technological information to the historian of arts and science. Over 2000 illustrations on 485 full page plates, most of them original size, show the trades and industries of a fascinating era in such great detail that the processes and shops might very well be reconstructed from them. The plates teem with life, with men, women, and children performing all of the thousands of operations necessary to the trades before and during the early stages of the industrial revolution. Plates are in sequence, and show general operations, closeups of difficult operations, and details of complex machinery. Such important and interesting trades and industries are illustrated as sowing, harvesting, beekeeping, cheesemaking, operating windmills, milling flour, charcoal burning, tobacco processing, indigo, fishing, arts of war, salt extraction, mining, smelting, casting iron, steel, extracting mercury, zinc, sulphur, copper, etc., slating, tinning, silverplating, gilding, making gunpowder, cannons, bells, shoeing horses, tanning, papermaking, printing, dyeing, and more than 40 other categories. Professor Gillispie, of Princeton, supplies a full commentary on all the plates, identifying operations, tools, processes, etc. This material, presented in a lively and lucid fashion, is of great interest to the reader interested in history of science and technology. Heavy library cloth. 920pp. 9 x 12.   T421 Two volume set **$18.50**

DE MAGNETE, William Gilbert. This classic work on magnetism founded a new science. Gilbert was the first to use the word "electricity", to recognize mass as distinct from weight, to discover the effect of heat on magnetic bodies; invent an electroscope, differentiate between static electricity and magnetism, conceive of the earth as a magnet. Written by the first great experimental scientist, this lively work is valuable not only as an historical landmark, but as the delightfully easy to follow record of a perpetually searching, ingenious mind. Translated by P. F. Mottelay. 25 page biographical memoir. 90 figures. lix + 368pp. 5⅜ x 8.   S470 Paperbound **$2.00**

A CONCISE HISTORY OF MATHEMATICS, D. Struik. Lucid study of development of mathematical ideas, techniques from Ancient Near East, Greece, Islamic science, Middle Ages, Renaissance, modern times. Important mathematicians are described in detail. Treatment is not anecdotal, but analytical development of ideas. "Rich in content, thoughtful in interpretation," U.S. QUARTERLY BOOKLIST. Non-technical; no mathematical training needed. Index. 60 illustrations, including Egyptian papyri, Greek mss., portraits of 31 eminent mathematicians. Bibliography. 2nd edition. xix + 299pp. 5⅜ x 8. **T255 Paperbound $1.75**

See also: **NON-EUCLIDEAN GEOMETRY, R. Bonola; THEORY OF DETERMINANTS IN HISTORICAL ORDER OF DEVELOPMENT, T. Muir; HISTORY OF THE THEORY OF ELASTICITY AND STRENGTH OF MATERIALS, I. Todhunter and K. Pearson; A SHORT HISTORY OF ASTRONOMY, A. Berry; CLASSICS OF SCIENCE.**

PHILOSOPHY OF SCIENCE AND MATHEMATICS

FOUNDATIONS OF SCIENCE: THE PHILOSOPHY OF THEORY AND EXPERIMENT, N. R. Campbell. A critique of the most fundamental concepts of science in general and physics in particular. Examines why certain propositions are accepted without question, demarcates science from philosophy, clarifies the understanding of the tools of science. Part One analyzes the presuppositions of scientific thought: existence of the material world, nature of scientific laws, multiplication of probabilities, etc.: Part Two covers the nature of experiment and the application of mathematics: conditions for measurement, relations between numerical laws and theories, laws of error, etc. An appendix covers problems arising from relativity, force, motion, space, and time. A classic in its field. Index. xiii + 565pp. 5⅝ x 8⅜.
S372 Paperbound $2.95

WHAT IS SCIENCE?, Norman Campbell. This excellent introduction explains scientific method, role of mathematics, types of scientific laws. Contents: 2 aspects of science, science & nature, laws of science, discovery of laws, explanation of laws, measurement & numerical laws, applications of science. 192pp. 5⅜ x 8. **S43 Paperbound $1.25**

THE VALUE OF SCIENCE, Henri Poincaré. Many of the most mature ideas of the "last scientific universalist" covered with charm and vigor for both the beginning student and the advanced worker. Discusses the nature of scientific truth, whether order is innate in the universe or imposed upon it by man, logical thought versus intuition (relating to math, through the works of Weierstrass, Lie, Klein, Riemann), time and space (relativity, psychological time, simultaneity), Hertz's concept of force, interrelationship of mathematical physics to pure math, values within disciplines of Maxwell, Carnot, Mayer, Newton, Lorentz, etc. Index. iii + 147pp. 5⅜ x 8. **S469 Paperbound $1.35**

SCIENCE AND METHOD, Henri Poincaré. Procedure of scientific discovery, methodology, experiment, idea-germination—the intellectual processes by which discoveries come into being. Most significant and most interesting aspects of development, application of ideas. Chapters cover selection of facts, chance, mathematical reasoning, mathematics, and logic; Whitehead, Russell, Cantor; the new mechanics, etc. 288pp. 5⅜ x 8. **S222 Paperbound $1.35**

SCIENCE AND HYPOTHESIS, Henri Poincaré. Creative psychology in science. How such concepts as number, magnitude, space, force, classical mechanics were developed, and how the modern scientist uses them in his thought. Hypothesis in physics, theories of modern physics. Introduction by Sir James Larmor. "Few mathematicians have had the breadth of vision of Poincaré, and none is his superior in the gift of clear exposition," E. T. Bell. Index. 272pp. 5⅜ x 8. **S221 Paperbound $1.35**

PHILOSOPHY AND THE PHYSICISTS, L. S. Stebbing. The philosophical aspects of modern science examined in terms of a lively critical attack on the ideas of Jeans and Eddington. Discusses the task of science, causality, determinism, probability, consciousness, the relation of the world of physics to that of everyday experience. Probes the philosophical significance of the Planck-Bohr concept of discontinuous energy levels, the inferences to be drawn from Heisenberg's Uncertainty Principle, the implications of "becoming" involved in the 2nd law of thermodynamics, and other problems posed by the discarding of Laplacean determinism. 285pp. 5⅜ x 8. **T480 Paperbound $1.65**

EXPERIMENT AND THEORY IN PHYSICS, Max Born. A Nobel laureate examines the nature and value of the counterclaims of experiment and theory in physics. Synthetic versus analytical scientific advances are analyzed in the work of Einstein, Bohr, Heisenberg, Planck, Eddington, Milne, and others by a fellow participant. 44pp. 5⅜ x 8. **S308 Paperbound 60¢**

THE NATURE OF PHYSICAL THEORY, P. W. Bridgman. Here is how modern physics looks to a highly unorthodox physicist—a Nobel laureate. Pointing out many absurdities of science, and demonstrating the inadequacies of various physical theories, Dr. Bridgman weighs and analyzes the contributions of Einstein, Bohr, Newton, Heisenberg, and many others. This is a non-technical consideration of the correlation of science and reality. Index. xi + 138pp. 5⅜ x 8.
S33 Paperbound **$1.25**

THE PHILOSOPHY OF SPACE AND TIME, H. Reichenbach. An important landmark in the development of the empiricist conception of geometry, covering the problem of the foundations of geometry, the theory of time, the consequences of Einstein's relativity, including: relations between theory and observations; coordinate and metrical properties of space; the psychological problem of visual intuition of non-Euclidean structures; and many other important topics in modern science and philosophy. The majority of ideas require only a knowledge of intermediate math. Introduction by R. Carnap. 49 figures. Index. xviii + 296pp. 5⅜ x 8.
S443 Paperbound **$2.00**

MATTER & MOTION, James Clerk Maxwell, This excellent exposition begins with simple particles and proceeds gradually to physical systems beyond complete analysis: motion, force, properties of centre of mass of material system, work, energy, gravitation, etc. Written with all Maxwell's original insights and clarity. Notes by E. Larmor. 17 diagrams. 178pp. 5⅜ x 8.
S188 Paperbound **$1.35**

THE ANALYSIS OF MATTER, Bertrand Russell. How do our senses concord with the new physics? This volume covers such topics as logical analysis of physics, prerelativity physics, causality, scientific inference, physics and perception, special and general relativity, Weyl's theory, tensors, invariants and their physical interpretation, periodicity and qualitative series. "The most thorough treatment of the subject that has yet been published," THE NATION. Introduction by L. E. Denonn. 422pp. 5⅜ x 8.
T231 Paperbound **$1.95**

SUBSTANCE AND FUNCTION, & EINSTEIN'S THEORY OF RELATIVITY, Ernst Cassirer. Two books bound as one. Cassirer establishes a philosophy of the exact sciences that takes into consideration newer developments in mathematics, and also shows historical connections. Partial contents: Aristotelian logic, Mill's analysis, Helmholtz & Kronecker, Russell & cardinal numbers, Euclidean vs. non-Euclidean geometry, Einstein's relativity. Bibliography. Index. xxi + 465pp. 5⅜ x 8.
T50 Paperbound **$2.00**

PRINCIPLES OF MECHANICS, Heinrich Hertz. This last work by the great 19th century physicist is not only a classic, but of great interest in the logic of science. Creating a new system of mechanics based upon space, time, and mass, it returns to axiomatic analysis, to understanding of the formal or structural aspects of science, taking into account logic, observation, and a priori elements. Of great historical importance to Poincaré, Carnap, Einstein, Milne. A 20-page introduction by R. S. Cohen, Wesleyan University, analyzes the implications of Hertz's thought and the logic of science. Bibliography. 13-page introduction by Helmholtz. xlii + 274pp. 5⅜ x 8.
S316 Clothbound **$3.50**
S317 Paperbound **$1.85**

THE PHILOSOPHICAL WRITINGS OF PEIRCE, edited by Justus Buchler. (Formerly published as THE PHILOSOPHY OF PEIRCE.) This is a carefully balanced exposition of Peirce's complete system, written by Peirce himself. It covers such matters as scientific method, pure chance vs. law, symbolic logic, theory of signs, pragmatism, experiment, and other topics. Introduction by Justus Buchler, Columbia University. xvi + 368pp. 5⅜ x 8.
T217 Paperbound **$1.95**

ESSAYS IN EXPERIMENTAL LOGIC, John Dewey. This stimulating series of essays touches upon the relationship between inquiry and experience, dependence of knowledge upon thought, character of logic; judgments of practice, data and meanings, stimuli of thought, etc. Index. viii + 444pp. 5⅜ x 8.
T73 Paperbound **$1.95**

LANGUAGE, TRUTH AND LOGIC, A. Ayer. A clear introduction to the Vienna and Cambridge schools of Logical Positivism. It sets up specific tests by which you can evaluate validity of ideas, etc. Contents: Function of philosophy, elimination of metaphysics, nature of analysis, a priori, truth and probability, etc. 10th printing. "I should like to have written it myself," Bertrand Russell. Index. 160pp. 5⅜ x 8.
T10 Paperbound **$1.25**

THE PSYCHOLOGY OF INVENTION IN THE MATHEMATICAL FIELD, J. Hadamard. Where do ideas come from? What role does the unconscious play? Are ideas best developed by mathematical reasoning, word reasoning, visualization? What are the methods used by Einstein, Poincaré, Galton, Riemann? How can these techniques be applied by others? Hadamard, one of the world's leading mathematicians, discusses these and other questions. xiii + 145pp. 5⅜ x 8.
T107 Paperbound **$1.25**

FOUNDATIONS OF GEOMETRY, Bertrand Russell. Analyzing basic problems in the overlap area between mathematics and philosophy, Nobel laureate Russell examines the nature of geometrical knowledge, the nature of geometry, and the application of geometry to space. It covers the history of non-Euclidean geometry, philosophic interpretations of geometry—especially Kant—projective and metrical geometry. This is most interesting as the solution offered in 1897 by a great mind to a problem still current. New introduction by Prof. Morris Kline of N. Y. University. xii + 201pp. 5⅜ x 8.
S232 Clothbound **$3.25**
S233 Paperbound **$1.60**

THE PRINCIPLES OF SCIENCE, A TREATISE ON LOGIC AND THE SCIENTIFIC METHOD, W. S. Jevons. Treating such topics as Inductive and Deductive Logic, the Theory of Number, Probability, and the Limits of Scientific Method, this milestone in the development of symbolic logic remains a stimulating contribution to the investigation of inferential validity in the natural and social sciences. It significantly advances Boole's logic, and contains a detailed introduction to the nature and methods of probability in physics, astronomy, everyday affairs, etc. In his introduction, Ernest Nagel of Columbia University says, "[Jevons] continues to be of interest as an attempt to articulate the logic of scientific inquiry." Index. liii + 786pp. 5⅜ x 8.   S446 Paperbound **$2.98**

Group theory, algebra, sets

LECTURES ON THE ICOSAHEDRON AND THE SOLUTION OF EQUATIONS OF THE FIFTH DEGREE, Felix Klein. The solution of quintics in terms of rotation of a regular icosahedron around its axes of symmetry. A classic & indispensable source for those interested in higher algebra, geometry, crystallography. Considerable explanatory material included. 230 footnotes, mostly bibliographic. 2nd edition, xvi + 289pp. 5⅜ x 8.   S314 Paperbound **$1.85**

LINEAR GROUPS, WITH AN EXPOSITION OF THE GALOIS FIELD THEORY, L. E. Dickson. The classic exposition of the theory of groups, well within the range of the graduate student. Part I contains the most extensive and thorough presentation of the theory of Galois Fields available, with a wealth of examples and theorems. Part II is a full discussion of linear groups of finite order. Much material in this work is based on Dickson's own contributions. Also includes expositions of Jordan, Lie, Abel, Betti-Mathieu, Hermite, etc. "A milestone in the development of modern algebra," W. Magnus, in his historical introduction to this edition. Index. xv + 312pp. 5⅜ x 8.   S482 Paperbound **$1.95**

INTRODUCTION TO THE THEORY OF GROUPS OF FINITE ORDER, R. Carmichael. Examines fundamental theorems and their application. Beginning with sets, systems, permutations, etc., it progresses in easy stages through important types of groups: Abelian, prime power, permutation, etc. Except 1 chapter where matrices are desirable, no higher math needed. 783 exercises, problems. Index. xvi + 447pp. 5⅜ x 8.   S299 Clothbound **$3.95**
                                                              S300 Paperbound **$2.00**

THEORY OF GROUPS OF FINITE ORDER, W. Burnside. First published some 40 years ago, this is still one of the clearest introductory texts. Partial contents: permutations, groups independent of representation, composition series of a group, isomorphism of a group with itself, Abelian groups, prime power groups, permutation groups, invariants of groups of linear substitution graphical representation, etc. 45pp. of notes. Indexes. xxiv + 512pp. 5⅜ x 8.
                                                              S38 Paperbound **$2.45**

THEORY AND APPLICATIONS OF FINITE GROUPS, G. A. Miller, H. F. Blichfeldt, L. E. Dickson. Unusually accurate and authoritative work, each section prepared by a leading specialist: Miller on substitution and abstract groups, Blichfeldt on finite groups of linear homogeneous transformations, Dickson on applications of finite groups. Unlike more modern works, this gives the concrete basis from which abstract group theory arose. Includes Abelian groups, prime-power groups, isomorphisms, matrix forms of linear transformations, Sylow groups, Galois' theory of algebraic equations, duplication of a cube, trisection of an angle, etc. 2 indexes. 267 problems. xvii + 390pp. 5⅜ x 8.   S216 Paperbound **$2.00**

CONTINUOUS GROUPS OF TRANSFORMATIONS, L. P. Eisenhart. Intensive study of the theory and geometrical applications of continuous groups of transformations; a standard work on the subject, called forth by the revolution in physics in the 1920's. Covers tensor analysis, Riemannian geometry, canonical parameters, transitivity, imprimitivity, differential invariants, the algebra of constants of structure, differential geometry, contact transformations, etc. "Likely to remain one of the standard works on the subject for many years . . . principal theorems are proved clearly and concisely, and the arrangement of the whole is coherent," MATHEMATICAL GAZETTE. Index. 72-item bibliography. 185 exercises. ix + 301pp. 5⅜ x 8.
                                                              S781 Paperbound **$1.85**

THE THEORY OF GROUPS AND QUANTUM MECHANICS, H. Weyl. Discussions of Schroedinger's wave equation, de Broglie's waves of a particle, Jordan-Hoelder theorem, Lie's continuous groups of transformations, Pauli exclusion principle, quantization of Maxwell-Dirac field equations, etc. Unitary geometry, quantum theory, groups, application of groups to quantum mechanics, symmetry permutation group, algebra of symmetric transformation, etc. 2nd revised edition. Bibliography. Index. xxii + 422pp. 5⅜ x 8.   S268 Clothbound **$4.50**
                                                              S269 Paperbound **$1.95**

ALGEBRAIC THEORIES, L. E. Dickson. Best thorough introduction to classical topics in higher algebra develops theories centering around matrices, invariants, groups. Higher algebra, Galois theory, finite linear groups, Klein's icosahedron, algebraic invariants, linear transformations, elementary divisors, invariant factors; quadratic, bi-linear, Hermitian forms, singly and in pairs. Proofs rigorous, detailed; topics developed lucidly, in close connection with their most frequent mathematical applications. Formerly "Modern Algebraic Theories." 155 problems. Bibliography. 2 indexes. 285pp. 5⅜ x 8.   S547 Paperbound **$1.50**

ALGEBRAS AND THEIR ARITHMETICS, L. E. Dickson. Provides the foundation and background necessary to any advanced undergraduate or graduate student studying abstract algebra. Begins with elementary introduction to linear transformations, matrices, field of complex numbers; proceeds to order, basal units, modulus, quaternions, etc.; develops calculus of linear sets, describes various examples of algebras including invariant, difference, nilpotent, semi-simple. "Makes the reader marvel at his genius for clear and profound analysis," Amer. Mathematical Monthly. Index. xii + 241pp. 5⅜ x 8. S616 Paperbound **$1.35**

THE THEORY OF EQUATIONS WITH AN INTRODUCTION TO THE THEORY OF BINARY ALGEBRAIC FORMS, W. S. Burnside and A. W. Panton. Extremely thorough and concrete discussion of the theory of equations, with extensive detailed treatment of many topics curtailed in later texts. Covers theory of algebraic equations, properties of polynomials, symmetric functions, derived functions, Horner's process, complex numbers and the complex variable, determinants and methods of elimination, invariant theory (nearly 100 pages), transformations, introduction to Galois theory, Abelian equations, and much more. Invaluable supplementary work for modern students and teachers. 759 examples and exercises. Index in each volume. Two volume set. Total of xxiv + 604pp. 5⅜ x 8.
S714 Vol I Paperbound **$1.85**
S715 Vol II Paperbound **$1.85**
The set **$3.70**

COMPUTATIONAL METHODS OF LINEAR ALGEBRA, V. N. Faddeeva, translated by **C. D. Benster.** First English translation of a unique and valuable work, the only work in English presenting a systematic exposition of the most important methods of linear algebra—classical and contemporary. Shows in detail how to derive numerical solutions of problems in mathematical physics which are frequently connected with those of linear algebra. Theory as well as individual practice. Part I surveys the mathematical background that is indispensable to what follows. Parts II and III, the conclusion, set forth the most important methods of solution, for both exact and iterative groups. One of the most outstanding and valuable features of this work is the 23 tables, double and triple checked for accuracy. These tables will not be found elsewhere. Author's preface. Translator's note. New bibliography and index. x + 252pp. 5⅜ x 8. S424 Paperbound **$1.95**

ALGEBRAIC EQUATIONS, E. Dehn. Careful and complete presentation of Galois' theory of algebraic equations; theories of Lagrange and Galois developed in logical rather than historical form, with a more thorough exposition than in most modern books. Many concrete applications and fully-worked-out examples. Discusses basic theory (very clear exposition of the symmetric group); isomorphic, transitive, and Abelian groups; applications of Lagrange's and Galois' theories; and much more. Newly revised by the author. Index. List of Theorems. xi + 208pp. 5⅜ x 8. S697 Paperbound **$1.45**

THEORY OF SETS, E. Kamke. Clearest, amplest introduction in English, well suited for independent study. Subdivision of main theory, such as theory of sets of points, are discussed, but emphasis is on general theory. Partial contents: rudiments of set theory, arbitrary sets and their cardinal numbers, ordered sets and their order types, well-ordered sets and their cardinal numbers. Bibliography. Key to symbols. Index. vii + 144pp. 5⅜ x 8.
S141 Paperbound **$1.35**

Number theory

INTRODUCTION TO THE THEORY OF NUMBERS, L. E. Dickson. Thorough, comprehensive approach with adequate coverage of classical literature, an introductory volume beginners can follow. Chapters on divisibility, congruences, quadratic residues & reciprocity, Diophantine equations, etc. Full treatment of binary quadratic forms without usual restriction to integral coefficients. Covers infinitude of primes, least residues, Fermat's theorem, Euler's phi function, Legendre's symbol, Gauss's lemma, automorphs, reduced forms, recent theorems of Thue & Siegel, many more. Much material not readily available elsewhere. 239 problems. Index. J figure. viii + 183pp. 5⅜ x 8. S342 Paperbound **$1.65**

ELEMENTS OF NUMBER THEORY, I. M. Vinogradov. Detailed 1st course for persons without advanced mathematics; 95% of this book can be understood by readers who have gone no farther than high school algebra. Partial contents: divisibility theory, important number theoretical functions, congruences, primitive roots and indices, etc. Solutions to both problems and exercises. Tables of primes, indices, etc. Covers almost every essential formula in elementary number theory! Translated from Russian. 233 problems, 104 exercises. viii + 227pp. 5⅜ x 8. S259 Paperbound **$1.60**

THEORY OF NUMBERS and DIOPHANTINE ANALYSIS, R. D. Carmichael. These two complete works in one volume form one of the most lucid introductions to number theory, requiring only a firm foundation in high school mathematics. "Theory of Numbers," partial contents: Eratosthenes' sieve, Euclid's fundamental theorem, G.C.F. and L.C.M. of two or more integers, linear congruences, etc "Diophantine Analysis": rational triangles, Pythagorean triangles, equations of third, fourth, higher degrees, method of functional equations, much more. "Theory of Numbers": 76 problems. Index. 94pp. "Diophantine Analysis": 222 problems. Index. 118pp. 5⅜ x 8. S529 Paperbound **$1.35**

CONTRIBUTIONS TO THE FOUNDING OF THE THEORY OF TRANSFINITE NUMBERS, Georg Cantor. These papers founded a new branch of mathematics. The famous articles of 1895-7 are translated, with an 82-page introduction by P. E. B. Jourdain dealing with Cantor, the background of his discoveries, their results, future possibilities. Bibliography. Index. Notes. ix + 211 pp. 5⅜ x 8. S45 Paperbound **$1.25**

See also: **TRANSCENDENTAL AND ALGEBRAIC NUMBERS, A. O. Gelfond.**

Probability theory and information theory

A PHILOSOPHICAL ESSAY ON PROBABILITIES, Marquis de Laplace. This famous essay explains without recourse to mathematics the principle of probability, and the application of probability to games of chance, natural philosophy, astronomy, many other fields. Translated from the 6th French edition by F. W. Truscott, F. L. Emory, with new introduction for this edition by E. T. Bell. 204pp. 5⅜ x 8. S166 Paperbound **$1.35**

MATHEMATICAL FOUNDATIONS OF INFORMATION THEORY, A. I. Khinchin. For the first time mathematicians, statisticians, physicists, cyberneticists, and communications engineers are offered a complete and exact introduction to this relatively new field. Entropy as a measure of a finite scheme, applications to coding theory, study of sources, channels and codes, detailed proofs of both Shannon theorems for any ergodic source and any stationary channel with finite memory, and much more are covered. Bibliography. vii + 120pp. 5⅜ x 8. S434 Paperbound **$1.35**

SELECTED PAPERS ON NOISE AND STOCHASTIC PROCESS, edited by **Prof. Nelson Wax,** U. of Illinois. 6 basic papers for newcomers in the field, for those whose work involves noise characteristics. Chandrasekhar, Uhlenbeck & Ornstein, Uhlenbeck & Ming, Rice, Doob. Included is Kac's Chauvenet-Prize winning Random Walk. Extensive bibliography lists 200 articles, up through 1953. 21 figures. 337pp. 6⅛ x 9¼. S262 Paperbound **$2.35**

THEORY OF PROBABILITY, William Burnside. Synthesis, expansion of individual papers presents numerous problems in classical probability, offering many original views succinctly, effectively. Game theory, cards, selections from groups; geometrical probability in such areas as suppositions as to probability of position of point on a line, points on surface of sphere, etc. Includes methods of approximation, theory of errors, direct calculation of probabilities, etc. Index. 136pp. 5⅜ x 8. S567 Paperbound **$1.00**

Vector and tensor analysis, matrix theory

VECTOR AND TENSOR ANALYSIS, A. P. Wills. Covers the entire field of vector and tensor analysis from elementary notions to dyads and non-Euclidean manifolds (especially detailed), absolute differentiation, the Lamé operator, the Riemann-Christoffel and Ricci-Einstein tensors, and the calculation of the Gaussian curvature of a surface. Many illustrations from electrical engineering, relativity theory, astro-physics, quantum mechanics. Presupposes only a good working knowledge of calculus. Exercises at end of each chapter. Intended for physicists and engineers as well as pure mathematicians. 44 diagrams. 114 problems. Bibliography. Index. xxxii + 285pp. 5⅜ x 8. S454 Paperbound **$1.75**

APPLICATIONS OF TENSOR ANALYSIS, A. J. McConnell. (Formerly APPLICATIONS OF THE ABSOLUTE DIFFERENTIAL CALCULUS.) An excellent text for understanding the application of tensor methods to familiar subjects such as dynamics, electricity, elasticity, and hydrodynamics. Explains the fundamental ideas and notation of tensor theory, the geometrical treatment of tensor algebra, the theory of differentiation of tensors, and includes a wealth of practical material. Bibliography. Index. 43 illustrations. 685 problems. xii + 381pp. 5⅜ x 8. S373 Paperbound **$1.85**

VECTOR AND TENSOR ANALYSIS, G. E. Hay. One of the clearest introductions to this increasingly important subject. Start with simple definitions, finish the book with a sure mastery of oriented Cartesian vectors, Christoffel symbols, solenoidal tensors, and their applications. Complete breakdown of plane, solid, analytical, differential geometry. Separate chapters on application. All fundamental formulae listed & demonstrated. 195 problems, 66 figures. viii + 193pp. 5⅜ x 8. S109 Paperbound **$1.75**

VECTOR ANALYSIS, FOUNDED UPON THE LECTURES OF J. WILLARD GIBBS, by E. B. Wilson. Still a first-rate introduction and supplementary text for students of mathematics and physics. Based on the pioneering lectures of Yale's great J. Willard Gibbs, can be followed by anyone who has had some calculus. Practical approach, stressing efficient use of combinations and functions of vectors. Worked examples from geometry, mechanics, hydrodynamics, gas theory, etc., as well as practice examples. Covers basic vector processes, differential and integral calculus in relation to vector functions, and theory of linear vector functions, forming an introduction to the study of multiple algebra and matrix theory. While the notation is not always modern, it is easily followed. xviii + 436pp. 5⅜ x 8. S656 Paperbound **$2.00**

PROBLEMS AND WORKED SOLUTIONS IN VECTOR ANALYSIS, L. R. Shorter. More pages of fully-worked-out examples than any other text on vector analysis. A self-contained course for home study or a fine classroom supplement. 138 problems and examples begin with fundamentals, then cover systems of coordinates, relative velocity and acceleration, the commutative and distributive laws, axial and polar vectors, finite displacements, the calculus of vectors, curl and divergence, etc. Final chapter treats applications in dynamics and physics: kinematics of a rigid body, equipotential surfaces, etc. "Very helpful . . . very comprehensive. A handy book like this . . . will fill a great want," MATHEMATICAL GAZETTE. Index. List of 174 important equations. 158 figures. xiv + 356pp. 5⅜ x 8. S135 Paperbound **$2.00**

THE THEORY OF DETERMINANTS, MATRICES, AND INVARIANTS, H. W. Turnbull. 3rd revised, corrected edition of this important study of virtually all the salient features and major theories of the subject. Covers Laplace identities, linear equations, differentiation, symbolic and direct methods for the reduction of invariants, seminvariants, Hilbert's Basis Theorem, Clebsch's Theorem, canonical forms, etc. New appendix contains a proof of Jacobi's lemma, further properties of symmetric determinants, etc. More than 350 problems. New references to recent developments. xviii + 374pp. 5⅜ x 8. S699 Paperbound **$2.00**

Differential equations, ordinary and partial, and integral equations

INTRODUCTION TO THE DIFFERENTIAL EQUATIONS OF PHYSICS, L. Hopf. Especially valuable to the engineer with no math beyond elementary calculus. Emphasizing intuitive rather than formal aspects of concepts, the author covers an extensive territory. Partial contents: Law of causality, energy theorem, damped oscillations, coupling by friction, cylindrical and spherical coordinates, heat source, etc. Index. 48 figures. 160pp. 5⅜ x 8.
S120 Paperbound **$1.25**

INTRODUCTION TO THE THEORY OF LINEAR DIFFERENTIAL EQUATIONS, E. G. Poole. Authoritative discussions of important topics, with methods of solution more detailed than usual, for students with background of elementary course in differential equations. Studies existence theorems, linearly independent solutions; equations with constant coefficients; with uniform analytic coefficients; regular singularities; the hypergeometric equation; conformal representation; etc. Exercises. Index. 210pp. 5⅜ x 8. S629 Paperbound **$1.65**

DIFFERENTIAL EQUATIONS FOR ENGINEERS, P. Franklin. Outgrowth of a course given 10 years at M. I. T. Makes most useful branch of pure math accessible for practical work. Theoretical basis of D.E.'s; solution of ordinary D.E.'s and partial derivatives arising from heat flow, steady-state temperature of a plate, wave equations; analytic functions; convergence of Fourier Series. 400 problems on electricity, vibratory systems, other topics. Formerly "Differential Equations for Electrical Engineers." Index. 41 illus. 307pp. 5⅜ x 8.
S601 Paperbound **$1.65**

DIFFERENTIAL EQUATIONS, F. R. Moulton. A detailed, rigorous exposition of all the non-elementary processes of solving ordinary differential equations. Several chapters devoted to the treatment of practical problems, especially those of a physical nature, which are far more advanced than problems usually given as illustrations. Includes analytic differential equations; variations of a parameter; integrals of differential equations; analytic implicit functions; problems of elliptic motion; sine-amplitude functions; deviation of formal bodies; Cauchy-Lipschitz process; linear differential equations with periodic coefficients; differential equations in infinitely many variations; much more. Historical notes. 10 figures. 222 problems. Index. xv + 395pp. 5⅜ x 8. S451 Paperbound **$2.00**

LECTURES ON CAUCHY'S PROBLEM, J. Hadamard. Based on lectures given at Columbia, Rome, this discusses work of Riemann, Kirchhoff, Volterra, and the author's own research on the hyperbolic case in linear partial differential equations. It extends spherical and cylindrical waves to apply to all (normal) hyperbolic equations. Partial contents: Cauchy's problem, fundamental formula, equations with odd number, with even number of independent variables; method of descent. 32 figures. Index. iii + 316pp. 5⅜ x 8. S105 Paperbound **$1.75**

PARTIAL DIFFERENTIAL EQUATIONS OF MATHEMATICAL PHYSICS, A. G. Webster. A keystone work in the library of every mature physicist, engineer, researcher. Valuable sections on elasticity, compression theory, potential theory, theory of sound, heat conduction, wave propagation, vibration theory. Contents include: deduction of differential equations, vibrations, normal functions, Fourier's series, Cauchy's method, boundary problems, method of Riemann-Volterra. Spherical, cylindrical, ellipsoidal harmonics, applications, etc. 97 figures. vii + 440pp. 5⅜ x 8. S263 Paperbound **$2.00**

ORDINARY DIFFERENTIAL EQUATIONS, E. L. Ince. A most compendious analysis in real and complex domains. Existence and nature of solutions, continuous transformation groups, solutions in an infinite form, definite integrals, algebraic theory, Sturmian theory, boundary problems, existence theorems, 1st order, higher order, etc. "Deserves the highest praise, a notable addition to mathematical literature," BULLETIN, AM. MATH. SOC. Historical appendix. Bibliography. 18 figures. viii + 558pp. 5⅜ x 8. S349 Paperbound **$2.55**

THEORY OF DIFFERENTIAL EQUATIONS, A. R. Forsyth. Out of print for over a decade, the complete 6 volumes (now bound as 3) of this monumental work represent the most comprehensive treatment of differential equations ever written. Historical presentation includes in 2500 pages every substantial development. Vol. 1, 2: EXACT EQUATIONS, PFAFF'S PROBLEM; ORDINARY EQUATIONS, NOT LINEAR: methods of Grassmann, Clebsch, Lie, Darboux; Cauchy's theorem; branch points; etc. Vol. 3, 4: ORDINARY EQUATIONS, NOT LINEAR; ORDINARY LINEAR EQUATIONS: Zeta Fuchsian functions, general theorems on algebraic integrals, Brun's theorem, equations with uniform periodic cofficients, etc. Vol. 4, 5: PARTIAL DIFFERENTIAL EQUATIONS: 2 existence-theorems, equations of theoretical dynamics, Laplace transformations, general transformation of equations of the 2nd order, much more. Indexes. Total of 2766pp. 5⅜ x 8. S576-7-8 Clothbound: the set **$15.00**

DIFFERENTIAL AND INTEGRAL EQUATIONS OF MECHANICS AND PHYSICS (DIE DIFFERENTIAL- UND INTEGRALGLEICHUNGEN DER MECHANIK UND PHYSIK), edited by P. Frank and R. von Mises. Most comprehensive and authoritative work on the mathematics of mathematical physics available today in the United States: the standard, definitive reference for teachers, physicists, engineers, and mathematicians—now published (in the original German) at a relatively inexpensive price for the first time! Every chapter in this 2,000-page set is by an expert in his field: Caratheodory, Courant, Frank, Mises, and a dozen others. Vol. I, on mathematics, gives concise but complete coverages of advanced calculus, differential equations, integral equations, and potential, and partial differential equations. Index. xxiii + 916pp. Vol. II (physics): classical mechanics, optics, continuous mechanics, heat conduction and diffusion, the stationary and quasi-stationary electromagnetic field, electromagnetic oscillations, and wave mechanics. Index. xxiv + 1106pp. Two volume set. Each volume available separately. 5⅝ x 8⅜. S787 Vol I Clothbound **$7.50**
S788 Vol II Clothbound **$7.50**
The set **$15.00**

MATHEMATICAL ANALYSIS OF ELECTRICAL AND OPTICAL WAVE-MOTION, Harry Bateman. Written by one of this century's most distinguished mathematical physicists, this is a practical introduction to those developments of Maxwell's electromagnetic theory which are directly connected with the solution of the partial differential equation of wave motion. Methods of solving wave-equation, polar-cylindrical coordinates, diffraction, transformation of coordinates, homogeneous solutions, electromagnetic fields with moving singularities, etc. Index. 168pp. 5⅜ x 8. S14 Paperbound **$1.60**

See also: **THE ANALYTICAL THEORY OF HEAT, J. Fourier; INTRODUCTION TO BESSEL FUNCTIONS, F. Bowman.**

Statistics

ELEMENTARY STATISTICS, WITH APPLICATIONS IN MEDICINE AND THE BIOLOGICAL SCIENCES, F. E. Croxton. A sound introduction to statistics for anyone in the physical sciences, assuming no prior acquaintance and requiring only a modest knowledge of math. All basic formulas carefully explained and illustrated; all necessary reference tables included. From basic terms and concepts, the study proceeds to frequency distribution, linear, non-linear, and multiple correlation, skewness, kurtosis, etc. A large section deals with reliability and significance of statistical methods. Containing concrete examples from medicine and biology, this book will prove unusually helpful to workers in those fields who increasingly must evaluate, check, and interpret statistics. Formerly titled "Elementary Statistics with Applications in Medicine." 101 charts. 57 tables. 14 appendices. Index. iv + 376pp. 5⅜ x 8. S506 Paperbound **$1.95**

METHODS OF STATISTICS, L. H. C. Tippett. A classic in its field, this unusually complete systematic introduction to statistical methods begins at beginner's level and progresses to advanced levels for experimenters and poll-takers in all fields of statistical research. Supplies fundamental knowledge of virtually all elementary methods in use today by sociologists, psychologists, biologists, engineers, mathematicians, etc. Explains logical and mathematical basis of each method described, with examples for each section. Covers frequency distributions and measures, inference from random samples, errors in large samples, simple analysis of variance, multiple and partial regression and correlation, etc. 4th revised (1952) edition. 16 charts. 5 significance tables. 152-item bibliography. 96 tables. 22 figures. 395pp. 6 x 9. S228 Clothbound **$7.50**

STATISTICS MANUAL, E. L. Crow, F. A. Davis, M. W. Maxfield. Comprehensive collection of classical, modern statistics methods, prepared under auspices of U. S. Naval Ordnance Test Station, China Lake, Calif. Many examples from ordnance will be valuable to workers in all fields. Emphasis is on use, with information on fiducial limits, sign tests, Chi-square runs, sensitivity, quality control, much more. "Well written . . . excellent reference work," Operations Research. Corrected edition of NAVORD Report 3360 NOTS 948. Introduction. Appendix of 32 tables, charts. Index. Bibliography. 95 illustrations. 306pp. 5⅜ x 8. S599 Paperbound **$1.55**

ANALYSIS & DESIGN OF EXPERIMENTS, H. B. Mann. Offers a method for grasping the analysis of variance and variance design within a short time. Partial contents: Chi-square distribution and analysis of variance distribution, matrices, quadratic forms, likelihood ration tests and tests of linear hypotheses, power of analysis, Galois fields, non-orthogonal data, interblock estimates, etc. 15pp. of useful tables. x + 195pp. 5 x 7⅜. S180 Paperbound **$1.45**

Numerical analysis, tables

PRACTICAL ANALYSIS, GRAPHICAL AND NUMERICAL METHODS, F. A. Willers. Translated by R. T. Beyer. Immensely practical handbook for engineers, showing how to interpolate, use various methods of numerical differentiation and integration, determine the roots of a single algebraic equation, system of linear equations, use empirical formulas, integrate differential equations, etc. Hundreds of shortcuts for arriving at numerical solutions. Special section on American calculating machines, by T. W. Simpson. 132 illustrations. 422pp. 5⅜ x 8.
S273 Paperbound **$2.00**

NUMERICAL SOLUTIONS OF DIFFERENTIAL EQUATIONS, H. Levy & E. A. Baggott. Comprehensive collection of methods for solving ordinary differential equations of first and higher order. All must pass 2 requirements: easy to grasp and practical, more rapid than school methods. Partial contents: graphical integration of differential equations, graphical methods for detailed solution. Numerical solution. Simultaneous equations and equations of 2nd and higher orders. "Should be in the hands of all in research in applied mathematics, teaching," NATURE. 21 figures. viii + 238pp. 5⅜ x 8. S168 Paperbound **$1.75**

NUMERICAL INTEGRATION OF DIFFERENTIAL EQUATIONS, Bennett, Milne & Bateman. Unabridged republication of original monograph prepared for National Research Council. New methods of integration of differential equations developed by 3 leading mathematicians: THE INTERPOLATIONAL POLYNOMIAL and SUCCESSIVE APPROXIMATIONS by A. A. Bennett; STEP-BY-STEP METHODS OF INTEGRATION by W. W. Milne; METHODS FOR PARTIAL DIFFERENTIAL EQUATIONS by H. Bateman. Methods for partial differential equations, transition from difference equations to differential equations, solution of differential equations to non-integral values of a parameter will interest mathematicians and physicists. 288 footnotes, mostly bibliographic; 235-item classified bibliography. 108pp. 5⅜ x 8. S305 Paperbound **$1.35**

INTRODUCTION TO RELAXATION METHODS, F. S. Shaw. Fluid mechanics, design of electrical networks, forces in structural frameworks, stress distribution, buckling, etc. Solve linear simultaneous equations, linear ordinary differential equations, partial differential equations, Eigen-value problems by relaxation methods. Detailed examples throughout. Special tables for dealing with awkwardly-shaped boundaries. Indexes. 253 diagrams. 72 tables. 400pp. 5⅜ x 8. S244 Paperbound **$2.45**

TABLES OF INDEFINITE INTEGRALS, G. Petit Bois. Comprehensive and accurate, this orderly grouping of over 2500 of the most useful indefinite integrals will save you hours of laborious mathematical groundwork. After a list of 49 common transformations of integral expressions, with a wide variety of examples, the book takes up algebraic functions, irrational monomials, products and quotients of binomials, transcendental functions, natural logs, etc. You will rarely or never encounter an integral of an algebraic or transcendental function not included here; any more comprehensive set of tables costs at least $12 or $15. Index. 2544 integrals. xii + 154pp. 6⅛ x 9¼. S225 Paperbound **$1.65**

A TABLE OF THE INCOMPLETE ELLIPTIC INTEGRAL OF THE THIRD KIND, R. G. Selfridge, J. E. Maxfield. The first complete 6 place tables of values of the incomplete integral of the third kind, prepared under the auspices of the Research Department of the U.S. Naval Ordnance Test Station. Calculated on an IBM type 704 calculator and thoroughly verified by echo-checking and a check integral at the completion of each value of **a.** Of inestimable value in problems where the surface area of geometrical bodies can only be expressed in terms of the incomplete integral of the third and lower kinds; problems in aero-, fluid-, and thermodynamics involving processes where nonsymmetrical repetitive volumes must be determined; various types of seismological problems; problems of magnetic potentials due to circular current; etc. Foreword. Acknowledgment. Introduction. Use of table. xiv + 805pp. 5⅝ x 8⅜. S501 Clothbound **$7.50**

MATHEMATICAL TABLES, H. B. Dwight. Unique for its coverage in one volume of almost every function of importance in applied mathematics, engineering, and the physical sciences. Three extremely fine tables of the three trig functions and their inverse functions to thousandths of radians; natural and common logarithms; squares, cubes; hyperbolic functions and the inverse hyperbolic functions; $(a^2 + b^2)$ exp. $\frac{1}{2}a$; complete elliptic integrals of the 1st and 2nd kind; sine and cosine integrals; exponential integrals $Ei(x)$ and $Ei(-x)$; binomial coefficients; factorials to 250; surface zonal harmonics and first derivatives; Bernoulli and Euler numbers and their logs to base of 10; Gamma function; normal probability integral; over 60 pages of Bessel functions; the Riemann Zeta function. Each table with formulae generally used, sources of more extensive tables, interpolation data, etc. Over half have columns of differences, to facilitate interpolation. Introduction. Index. viii + 231pp. 5⅜ x 8.
S445 Paperbound **$1.75**

TABLES OF FUNCTIONS WITH FORMULAE AND CURVES, E. Jahnke & F. Emde. The world's most comprehensive 1-volume English-text collection of tables, formulae, curves of transcendent functions. 4th corrected edition, new 76-page section giving tables, formulae for elementary functions—not in other English editions. Partial contents: sine, cosine, logarithmic integral; factorial function; error integral; theta functions; elliptic integrals, functions; Legendre, Bessel, Riemann, Mathieu, hypergeometric functions, etc. Supplementary books. Bibliography. Indexed. "Out of the way functions for which we know no other source," SCIENTIFIC COMPUTING SERVICE, Ltd. 212 figures. 400pp. 5⅜ x 8.   S133 Paperbound **$2.00**

JACOBIAN ELLIPTIC FUNCTION TABLES, L. M. Milne-Thomson. An easy to follow, practical book which gives not only useful numerical tables, but also a complete elementary sketch of the application of elliptic functions. It covers Jacobian elliptic functions and a description of their principal properties; complete elliptic integrals; Fourier series and power series expansions; periods, zeros, poles, residues, formulas for special values of the argument; transformations, approximations, elliptic integrals, conformal mapping, factorization of cubic and quartic polynomials; application to the pendulum problem; etc. Tables and graphs form the body of the book: Graph, 5 figure table of the elliptic function sn (u m); cn (u m); dn (u m). ·8 figure table of complete elliptic integrals K, K', E, E', and the nome q. 7 figure table of the Jacobian zeta-function Z(u). 3 figures. xi + 123pp. 5⅜ x 8.
S194 Paperbound **$1.35**

PHYSICS

General physics

FOUNDATIONS OF PHYSICS, R. B. Lindsay & H. Margenau. Excellent bridge between semi-popular works & technical treatises. A discussion of methods of physical description, construction of theory; valuable for physicist with elementary calculus who is interested in ideas that give meaning to data, tools of modern physics. Contents include symbolism, mathematical equations; space & time foundations of mechanics; probability; physics & continua; electron theory; special & general relativity; quantum mechanics; causality. "Thorough and yet not overdetailed. Unreservedly recommended," NATURE (London). Unabridged, corrected edition. List of recommended readings. 35 illustrations. xi + 537pp. 5⅜ x 8.
S377 Paperbound **$2.45**

FUNDAMENTAL FORMULAS OF PHYSICS, ed. by D. H. Menzel. Highly useful, fully inexpensive reference and study text, ranging from simple to highly sophisticated operations. Mathematics integrated into text—each chapter stands as short textbook of field represented. Vol. 1: Statistics, Physical Constants, Special Theory of Relativity, Hydrodynamics, Aerodynamics, Boundary Value Problems in Math. Physics; Viscosity, Electromagnetic Theory, etc. Vol. 2: Sound, Acoustics, Geometrical Optics, Electron Optics, High-Energy Phenomena, Magnetism, Biophysics, much more. Index. Total of 800pp. 5⅜ x 8.   Vol. 1 S595 Paperbound **$2.00**
Vol. 2 S596 Paperbound **$2.00**

MATHEMATICAL PHYSICS, D. H. Menzel. Thorough one-volume treatment of the mathematical techniques vital for classic mechanics, electromagnetic theory, quantum theory, and relativity. Written by the Harvard Professor of Astrophysics for junior, senior, and graduate courses, it gives clear explanations of all those aspects of function theory, vectors, matrices, dyadics, tensors, partial differential equations, etc., necessary for the understanding of the various physical theories. Electron theory, relativity, and other topics seldom presented appear here in considerable detail. Scores of definitions, conversion factors, dimensional constants, etc. "More detailed than normal for an advanced text . . . excellent set of sections on Dyadics, Matrices, and Tensors," JOURNAL OF THE FRANKLIN INSTITUTE. Index. 193 problems, with answers. x + 412pp. 5⅜ x 8.   S56 Paperbound **$2.00**

THE SCIENTIFIC PAPERS OF J. WILLARD GIBBS. All the published papers of America's outstanding theoretical scientist (except for "Statistical Mechanics" and "Vector Analysis"). Vol I (thermodynamics) contains one of the most brilliant of all 19th-century scientific papers—the 300-page "On the Equilibrium of Heterogeneous Substances," which founded the science of physical chemistry, and clearly stated a number of highly important natural laws for the first time; 8 other papers complete the first volume. Vol II includes 2 papers on dynamics, 8 on vector analysis and multiple algebra, 5 on the electromagnetic theory of light, and 6 miscellaneous papers. Biographical sketch by H. A. Bumstead. Total of xxxvi + 718pp. 5⅝ x 8⅜.
S721 Vol I Paperbound **$2.00**
S722 Vol II Paperbound **$2.00**
The set **$4.00**

Relativity, quantum theory, nuclear physics

THE PRINCIPLE OF RELATIVITY, A. Einstein, H. Lorentz, M. Minkowski, H. Weyl. These are the 11 basic papers that founded the general and special theories of relativity, all translated into English. Two papers by Lorentz on the Michelson experiment, electromagnetic phenomena. Minkowski's SPACE & TIME, and Weyl's GRAVITATION & ELECTRICITY. 7 epoch-making papers by Einstein: ELECTROMAGNETICS OF MOVING BODIES, INFLUENCE OF GRAVITATION IN PROPAGATION OF LIGHT, COSMOLOGICAL CONSIDERATIONS, GENERAL THEORY, and 3 others. 7 diagrams. Special notes by A. Sommerfeld. 224pp. 5⅜ x 8.
S81 Paperbound **$1.75**

SPACE TIME MATTER, Hermann Weyl. "The standard treatise on the general theory of relativity," (Nature), written by a world-renowned scientist, provides a deep clear discussion of the logical coherence of the general theory, with introduction to all the mathematical tools needed: Maxwell, analytical geometry, non-Euclidean geometry, tensor calculus, etc. Basis is classical space-time, before absorption of relativity. Partial contents: Euclidean space, mathematical form, metrical continuum, relativity of time and space, general theory. 15 diagrams. Bibliography. New preface for this edition. xviii + 330pp. 5⅜ x 8.
S267 Paperbound **$1.85**

PRINCIPLES OF QUANTUM MECHANICS, W. V. Houston. Enables student with working knowledge of elementary mathematical physics to develop facility in use of quantum mechanics, understand published work in field. Formulates quantum mechanics in terms of Schroedinger's wave mechanics. Studies evidence for quantum theory, for inadequacy of classical mechanics, 2 postulates of quantum mechanics; numerous important, fruitful applications of quantum mechanics in spectroscopy, collision problems, electrons in solids; other topics. "One of the most rewarding features . . . is the interlacing of problems with text," Amer. J. of Physics. Corrected edition. 21 illus. Index. 296pp. 5⅜ x 8.   S524 Paperbound **$1.85**

PHYSICAL PRINCIPLES OF THE QUANTUM THEORY, Werner Heisenberg. A Nobel laureate discusses quantum theory; Heisenberg's own work, Compton, Schroedinger, Wilson, Einstein, many others. Written for physicists, chemists who are not specialists in quantum theory, only elementary formulae are considered in the text; there is a mathematical appendix for specialists. Profound without sacrifice of clarity. Translated by C. Eckart, F. Hoyt. 18 figures. 192pp. 5⅜ x 8.
S113 Paperbound **$1.25**

SELECTED PAPERS ON QUANTUM ELECTRODYNAMICS, edited by **J. Schwinger.** Facsimiles of papers which established quantum electrodynamics, from initial successes through today's position as part of the larger theory of elementary particles. First book publication in any language of these collected papers of Bethe, Bloch, Dirac, Dyson, Fermi, Feynman, Heisenberg, Kusch, Lamb, Oppenheimer, Pauli, Schwinger, Tomonoga, Weisskopf, Wigner, etc. 34 papers in all, 29 in English, 1 in French, 3 in German, 1 in Italian. Preface and historical commentary by the editor. xvii + 423pp. 6⅛ x 9¼.
S444 Paperbound **$2.45**

THE FUNDAMENTAL PRINCIPLES OF QUANTUM MECHANICS, WITH ELEMENTARY APPLICATIONS, E. C. Kemble. An inductive presentation, for the graduate student or specialist in some other branch of physics. Assumes some acquaintance with advanced math; apparatus necessary beyond differential equations and advanced calculus is developed as needed. Although a general exposition of principles, hundreds of individual problems are fully treated, with applications of theory being interwoven with development of the mathematical structure. The author is the Professor of Physics at Harvard Univ. "This excellent book would be of great value to every student . . . a rigorous and detailed mathematical discussion of all of the principal quantum-mechanical methods . . . has succeeded in keeping his presentations clear and understandable," Dr. Linus Pauling, J. of the American Chemical Society. Appendices: calculus of variations, math. notes, etc. Indexes. 611pp. 5⅜ x 8.
S472 Paperbound **$2.95**

ATOMIC SPECTRA AND ATOMIC STRUCTURE, G. Herzberg. Excellent general survey for chemists, physicists specializing in other fields. Partial contents: simplest line spectra and elements of atomic theory, building-up principle and periodic system of elements, hyperfine structure of spectral lines, some experiments and applications. Bibliography. 80 figures. Index. xii + 257pp. 5⅜ x 8.
S115 Paperbound **$1.95**

THE THEORY AND THE PROPERTIES OF METALS AND ALLOYS, N. F. Mott, H. Jones. Quantum methods used to develop mathematical models which show interrelationship of basic chemical phenomena with crystal structure, magnetic susceptibility, electrical, optical properties. Examines thermal properties of crystal lattice, electron motion in applied field, cohesion, electrical resistance, noble metals, para-, dia-, and ferromagnetism, etc. "Exposition . . . clear . . . mathematical treatment . . . simple," Nature. 138 figures. Bibliography. Index. xiii + 320pp. 5⅜ x 8.
S456 Paperbound **$1.85**

FOUNDATIONS OF NUCLEAR PHYSICS, edited by **R. T. Beyer.** 13 of the most important papers on nuclear physics reproduced in facsimile in the original languages of their authors: the papers most often cited in footnotes, bibliographies. Anderson, Curie, Joliot, Chadwick, Fermi, Lawrence, Cockcroft, Hahn, Yukawa. UNPARALLELED BIBLIOGRAPHY. 122 double-columned pages, over 4,000 articles, books classified. 57 figures. 288pp. 6⅛ x 9¼.
S19 Paperbound **$1.75**

MESON PHYSICS, R. E. Marshak. Traces the basic theory, and explicitly presents results of experiments with particular emphasis on theoretical significance. Phenomena involving mesons as virtual transitions are avoided, eliminating some of the least satisfactory predictions of meson theory. Includes production and study of π mesons at nonrelativistic nucleon energies, contrasts between π and μ mesons, phenomena associated with nuclear interaction of π mesons, etc. Presents early evidence for new classes of particles and indicates theoretical difficulties created by discovery of heavy mesons and hyperons. Name and subject indices. Unabridged reprint. viii + 378pp. 5⅜ x 8.   S500 Paperbound **$1.95**

See also: **STRANGE STORY OF THE QUANTUM, B. Hoffmann; FROM EUCLID TO EDDINGTON, E. Whittaker; MATTER AND LIGHT, THE NEW PHYSICS, L. de Broglie; THE EVOLUTION OF SCIENTIFIC THOUGHT FROM NEWTON TO EINSTEIN, A. d'Abro; THE RISE OF THE NEW PHYSICS, A. d'Abro; THE THEORY OF GROUPS AND QUANTUM MECHANICS, H. Weyl; SUBSTANCE AND FUNCTION, & EINSTEIN'S THEORY OF RELATIVITY, E. Cassirer; FUNDAMENTAL FORMULAS OF PHYSICS, D. H. Menzel.**

Hydrodynamics

HYDRODYNAMICS, H. Dryden, F. Murnaghan, Harry Bateman. Published by the National Research Council in 1932 this enormous volume offers a complete coverage of classical hydrodynamics. Encyclopedic in quality. Partial contents: physics of fluids, motion, turbulent flow, compressible fluids, motion in 1, 2, 3 dimensions; viscous fluids rotating, laminar motion, resistance of motion through viscous fluid, eddy viscosity, hydraulic flow in channels of various shapes, discharge of gases, flow past obstacles, etc. Bibliography of over 2,900 items. Indexes. 23 figures. 634pp. 5⅜ x 8.   S303 Paperbound **$2.75**

A TREATISE ON HYDRODYNAMICS, A. B. Basset. Favorite text on hydrodynamics for 2 generations of physicists, hydrodynamical engineers, oceanographers, ship designers, etc. Clear enough for the beginning student, and thorough source for graduate students and engineers on the work of d'Alembert, Euler, Laplace, Lagrange, Poisson, Green, Clebsch, Stokes, Cauchy, Helmholtz, J. J. Thomson, Love, Hicks, Greenhill, Besant, Lamb, etc. Great amount of documentation on entire theory of classical hydrodynamics. Vol I: theory of motion of frictionless liquids, vortex, and cyclic irrotational motion, etc. 132 exercises. Bibliography. 3 Appendixes. xii + 264pp. Vol II: motion in viscous liquids, harmonic analysis, theory of tides, etc. 112 exercises. Bibliography. 4 Appendixes. xv + 328pp. Two volume set. 5⅜ x 8.
S724 Vol I Paperbound **$1.75**
S725 Vol II Paperbound **$1.75**
The set **$3.50**

HYDRODYNAMICS, Horace Lamb. Internationally famous complete coverage of standard reference work on dynamics of liquids & gases. Fundamental theorems, equations, methods, solutions, background, for classical hydrodynamics. Chapters include Equations of Motion, Integration of Equations in Special Gases, Irrotational Motion, Motion of Liquid in 2 Dimensions, Motion of Solids through Liquid-Dynamical Theory, Vortex Motion, Tidal Waves, Surface Waves, Waves of Expansion, Viscosity, Rotating Masses of liquids. Excellently planned, arranged; clear, lucid presentation. 6th enlarged, revised edition. Index. Over 900 footnotes, mostly bibliographical. 119 figures. xv + 738pp. 6⅛ x 9¼.   S256 Paperbound **$2.95**

See also: **FUNDAMENTAL FORMULAS OF PHYSICS, D. H. Menzel; THEORY OF FLIGHT, R. von Mises; FUNDAMENTALS OF HYDRO- AND AEROMECHANICS, L. Prandtl and O. G. Tietjens; APPLIED HYDRO- AND AEROMECHANICS, L. Prandtl and O. G. Tietjens; HYDRAULICS AND ITS APPLICATIONS, A. H. Gibson; FLUID MECHANICS FOR HYDRAULIC ENGINEERS, H. Rouse.**

Acoustics, optics, electromagnetics

ON THE SENSATIONS OF TONE, Hermann Helmholtz. This is an unmatched coordination of such fields as acoustical physics, physiology, experiment, history of music. It covers the entire gamut of musical tone. Partial contents: relation of musical science to acoustics, physical vs. physiological acoustics, composition of vibration, resonance, analysis of tones by sympathetic resonance, beats, chords, tonality, consonant chords, discords, progression of parts, etc. 33 appendixes discuss various aspects of sound, physics, acoustics, music, etc. Translated by A. J. Ellis. New introduction by Prof. Henry Margenau of Yale. 68 figures. 43 musical passages analyzed. Over 100 tables. Index. xix + 576pp. 6⅛ x 9¼.
S114 Paperbound **$2.95**

THE THEORY OF SOUND, Lord Rayleigh. Most vibrating systems likely to be encountered in practice can be tackled successfully by the methods set forth by the great Nobel laureate, Lord Rayleigh. Complete coverage of experimental, mathematical aspects of sound theory. Partial contents: Harmonic motions, vibrating systems in general, lateral vibrations of bars, curved plates or shells, applications of Laplace's functions to acoustical problems, fluid friction, plane vortex-sheet, vibrations of solid bodies, etc. This is the first inexpensive edition of this great reference and study work. Bibliography. Historical introduction by R. B. Lindsay. Total of 1040pp. 97 figures. 5⅜ x 8.
S292, S293, Two volume set, paperbound, **$4.00**

THE DYNAMICAL THEORY OF SOUND, H. Lamb. Comprehensive mathematical treatment of the physical aspects of sound, covering the theory of vibrations, the general theory of sound, and the equations of motion of strings, bars, membranes, pipes, and resonators. Includes chapters on plane, spherical, and simple harmonic waves, and the Helmholtz Theory of Audition. Complete and self-contained development for student and specialist; all fundamental differential equations solved completely. Specific mathematical details for such important phenomena as harmonics, normal modes, forced vibrations of strings, theory of reed pipes, etc. Index. Bibliography. 86 diagrams. viii + 307pp. 5⅜ x 8.
S655 Paperbound **$1.50**

WAVE PROPAGATION IN PERIODIC STRUCTURES, L. Brillouin. A general method and application to different problems: pure physics, such as scattering of X-rays of crystals, thermal vibration in crystal lattices, electronic motion in metals; and also problems of electrical engineering. Partial contents: elastic waves in 1-dimensional lattices of point masses. Propagation of waves along 1-dimensional lattices. Energy flow. 2 dimensional, 3 dimensional lattices. Mathieu's equation. Matrices and propagation of waves along an electric line. Continuous electric lines. 131 illustrations. Bibliography. Index. xii + 253pp. 5⅜ x 8.
S34 Paperbound **$1.85**

THEORY OF VIBRATIONS, N. W. McLachlan. Based on an exceptionally successful graduate course given at Brown University, this discusses linear systems having 1 degree of freedom, forced vibrations of simple linear systems, vibration of flexible strings, transverse vibrations of bars and tubes, transverse vibration of circular plate, sound waves of finite amplitude, etc. Index. 99 diagrams. 160pp. 5⅜ x 8.
S190 Paperbound **$1.35**

LOUD SPEAKERS: THEORY, PERFORMANCE, TESTING AND DESIGN, N. W. McLachlan. Most comprehensive coverage of theory, practice of loud speaker design, testing; classic reference, study manual in field. First 12 chapters deal with theory, for readers mainly concerned with math. aspects; last 7 chapters will interest reader concerned with testing, design. Partial contents: principles of sound propagation, fluid pressure on vibrators, theory of moving-coil principle, transients, driving mechanisms, response curves, design of horn type moving coil speakers, electrostatic speakers, much more. Appendix. Bibliography. Index. 165 illustrations, charts. 411pp. 5⅜ x 8.
S588 Paperbound **$2.25**

MICROWAVE TRANSMISSION, J. S. Slater. First text dealing exclusively with microwaves, brings together points of view of field, circuit theory, for graduate student in physics, electrical engineering, microwave technician. Offers valuable point of view not in most later studies. Uses Maxwell's equations to study electromagnetic field, important in this area. Partial contents: infinite line with distributed parameters, impedance of terminated line, plane waves, reflections, wave guides, coaxial line, composite transmission lines, impedance matching, etc. Introduction. Index. 76 illus. 319pp. 5⅜ x 8.
S564 Paperbound **$1.50**

THE ANALYSIS OF SENSATIONS, Ernst Mach. Great study of physiology, psychology of perception, shows Mach's ability to see material freshly, his "incorruptible skepticism and independence." (Einstein). Relation of problems of psychological perception to classical physics, supposed dualism of physical and mental, principle of continuity, evolution of senses, will as organic manifestation, scores of experiments, observations in optics, acoustics, music, graphics, etc. New introduction by T. S. Szasz, M. D. 58 illus. 300-item bibliography. Index. 404pp. 5⅜ x 8.
S525 Paperbound **$1.75**

APPLIED OPTICS AND OPTICAL DESIGN, A. E. Conrady. With publication of vol. 2, standard work for designers in optics is now complete for first time. Only work of its kind in English; only detailed work for practical designer and self-taught. Requires, for bulk of work, no math above trig. Step-by-step exposition, from fundamental concepts of geometrical, physical optics, to systematic study, design, of almost all types of optical systems. Vol. 1: all ordinary ray-tracing methods; primary aberrations; necessary higher aberration for design of telescopes, low-power microscopes, photographic equipment. Vol. 2: (Completed from author's notes by R. Kingslake, Dir. Optical Design, Eastman Kodak.) Special attention to high-power microscope, anastigmatic photographic objectives. "An indispensable work," J., Optical Soc. of Amer. "As a practical guide this book has no rival," Transactions, Optical Soc. Index. Bibliography. 193 diagrams. 852pp. 6⅛ x 9¼.
Vol. 1 T611 Paperbound **$2.95**
Vol. 2 T612 Paperbound **$2.95**

THE THEORY OF OPTICS, Paul Drude. One of finest fundamental texts in physical optics, classic offers thorough coverage, complete mathematical treatment of basic ideas. Includes fullest treatment of application of thermodynamics to optics; sine law in formation of images, transparent crystals, magnetically active substances, velocity of light, apertures, effects depending upon them, polarization, optical instruments, etc. Introduction by A. A. Michelson. Index. 110 illus. 567pp. 5⅜ x 8.
S532 Paperbound **$2.45**

OPTICKS, Sir Isaac Newton. In its discussions of light, reflection, color, refraction, theories of wave and corpuscular theories of light, this work is packed with scores of insights and discoveries. In its precise and practical discussion of construction of optical apparatus, contemporary understandings of phenomena it is truly fascinating to modern physicists, astronomers, mathematicians. Foreword by Albert Einstein. Preface by I. B. Cohen of Harvard University. 7 pages of portraits, facsimile pages, letters, etc. cxvi + 414pp. 5⅜ x 8.
S205 Paperbound **$2.00**

OPTICS AND OPTICAL INSTRUMENTS: AN INTRODUCTION WITH SPECIAL REFERENCE TO PRACTICAL APPLICATIONS, B. K. Johnson. An invaluable guide to basic practical applications of optical principles, which shows how to set up inexpensive working models of each of the four main types of optical instruments—telescopes, microscopes, photographic lenses, optical projecting systems. Explains in detail the most important experiments for determining their accuracy, resolving power, angular field of view, amounts of aberration, all other necessary facts about the instruments. Formerly "Practical Optics." Index. 234 diagrams. Appendix. 224pp. 5⅜ x 8.
S642 Paperbound **$1.65**

PRINCIPLES OF PHYSICAL OPTICS, Ernst Mach. This classical examination of the propagation of light, color, polarization, etc. offers an historical and philosophical treatment that has never been surpassed for breadth and easy readability. Contents: Rectilinear propagation of light. Reflection, refraction. Early knowledge of vision. Dioptrics. Composition of light. Theory of color and dispersion. Periodicity. Theory of interference. Polarization. Mathematical representation of properties of light. Propagation of waves, etc. 279 illustrations, 10 portraits. Appendix. Indexes. 324pp. 5⅜ x 8.
S178 Paperbound **$1.75**

FUNDAMENTALS OF ELECTRICITY AND MAGNETISM, L. B. Loeb. For students of physics, chemistry, or engineering who want an introduction to electricity and magnetism on a higher level and in more detail than general elementary physics texts provide. Only elementary differential and integral calculus is assumed. Physical laws developed logically, from magnetism to electric currents, Ohm's law, electrolysis, and on to static electricity, induction, etc. Covers an unusual amount of material; one third of book on modern material: solution of wave equation, photoelectric and thermionic effects, etc. Complete statement of the various electrical systems of units and interrelations. 2 Indexes. 75 pages of problems with answers stated. Over 300 figures and diagrams. xix +669pp. 5⅜ x 8.
S745 Paperbound **$2.75**

THE ELECTROMAGNETIC FIELD, Max Mason & Warren Weaver. Used constantly by graduate engineers. Vector methods exclusively: detailed treatment of electrostatics, expansion methods, with tables converting any quantity into absolute electromagnetic, absolute electrostatic, practical units. Discrete charges, ponderable bodies, Maxwell field equations, etc. Introduction. Indexes. 416pp. 5⅜ x 8.
S185 Paperbound **$2.00**

ELECTRICAL THEORY ON THE GIORGI SYSTEM, P. Cornelius. A new clarification of the fundamental concepts of electricity and magnetism, advocating the convenient m.k.s. system of units that is steadily gaining followers in the sciences. Illustrating the use and effectiveness of his terminology with numerous applications to concrete technical problems, the author here expounds the famous Giorgi system of electrical physics. His lucid presentation and well-reasoned, cogent argument for the universal adoption of this system form one of the finest pieces of scientific exposition in recent years. 28 figures. Index. Conversion tables for translating earlier data into modern units. Translated from 3rd Dutch edition by L. J. Jolley. x + 187pp. 5½ x 8¾.
S909 Clothbound **$6.00**

THEORY OF ELECTRONS AND ITS APPLICATION TO THE PHENOMENA OF LIGHT AND RADIANT HEAT, H. Lorentz. Lectures delivered at Columbia University by Nobel laureate Lorentz. Unabridged, they form a historical coverage of the theory of free electrons, motion, absorption of heat, Zeeman effect, propagation of light in molecular bodies, inverse Zeeman effect, optical phenomena in moving bodies, etc. 109 pages of notes explain the more advanced sections. Index. 9 figures. 352pp. 5⅜ x 8.
S173 Paperbound **$1.85**

TREATISE ON ELECTRICITY AND MAGNETISM, James Clerk Maxwell. For more than 80 years a seemingly inexhaustible source of leads for physicists, mathematicians, engineers. Total of 1082pp. on such topics as Measurement of Quantities, Electrostatics, Elementary Mathematical Theory of Electricity, Electrical Work and Energy in a System of Conductors, General Theorems, Theory of Electrical Images, Electrolysis, Conduction, Polarization, Dielectrics, Resistance, etc. "The greatest mathematical physicist since Newton," Sir James Jeans. 3rd edition. 107 figures, 21 plates. 1082pp. 5⅜ x 8.
S636-7, 2 volume set, paperbound **$4.00**

See also: **FUNDAMENTAL FORMULAS OF PHYSICS, D. H. Menzel; MATHEMATICAL ANALYSIS OF ELECTRICAL & OPTICAL WAVE MOTION, H. Bateman.**

Mechanics, dynamics, thermodynamics, elasticity

MECHANICS VIA THE CALCULUS, P. W. Norris, W. S. Legge. Covers almost everything, from linear motion to vector analysis: equations determining motion, linear methods, compounding of simple harmonic motions, Newton's laws of motion, Hooke's law, the simple pendulum, motion of a particle in 1 plane, centers of gravity, virtual work, friction, kinetic energy of rotating bodies, equilibrium of strings, hydrostatics, sheering stresses, elasticity, etc. 550 problems. 3rd revised edition. xii + 367pp. 6 x 9.
S207 Clothbound **$3.95**

MECHANICS, J. P. Den Hartog. Already a classic among introductory texts, the M.I.T. professor's lively and discursive presentation is equally valuable as a beginner's text, an engineering student's refresher, or a practicing engineer's reference. Emphasis in this highly readable text is on illuminating fundamental principles and showing how they are embodied in a great number of real engineering and design problems: trusses, loaded cables, beams, jacks, hoists, etc. Provides advanced material on relative motion and gyroscopes not usual in introductory texts. "Very thoroughly recommended to all those anxious to improve their real understanding of the principles of mechanics." MECHANICAL WORLD. Index. List of equations. 334 problems, all with answers. Over 550 diagrams and drawings. ix + 462pp. 5⅜ x 8.
S754 Paperbound **$2.00**

THEORETICAL MECHANICS: AN INTRODUCTION TO MATHEMATICAL PHYSICS, J. S. Ames, F. D. Murnaghan. A mathematically rigorous development of theoretical mechanics for the advanced student, with constant practical applications. Used in hundreds of advanced courses. An unusually thorough coverage of gyroscopic and baryscopic material, detailed analyses of the Corilis acceleration, applications of Lagrange's equations, motion of the double pendulum, Hamilton-Jacobi partial differential equations, group velocity and dispersion, etc. Special relativity is also included. 159 problems. 44 figures. ix + 462pp. 5⅜ x 8.
S461 Paperbound **$2.00**

THEORETICAL MECHANICS: STATICS AND THE DYNAMICS OF A PARTICLE, W. D. MacMillan. Used for over 3 decades as a self-contained and extremely comprehensive advanced undergraduate text in mathematical physics, physics, astronomy, and deeper foundations of engineering. Early sections require only a knowledge of geometry; later, a working knowledge of calculus. Hundreds of basic problems, including projectiles to the moon, escape velocity, harmonic motion, ballistics, falling bodies, transmission of power, stress and strain, elasticity, astronomical problems. 340 practice problems plus many fully worked out examples make it possible to test and extend principles developed in the text. 200 figures. xvii + 430pp. 5⅜ x 8.
S467 Paperbound **$2.00**

THEORETICAL MECHANICS: THE THEORY OF THE POTENTIAL, W. D. MacMillan. A comprehensive, well balanced presentation of potential theory, serving both as an introduction and a reference work with regard to specific problems, for physicists and mathematicians. No prior knowledge of integral relations is assumed, and all mathematical material is developed as it becomes necessary. Includes: Attraction of Finite Bodies; Newtonian Potential Function; Vector Fields, Green and Gauss Theorems; Attractions of Surfaces and Lines; Surface Distribution of Matter; Two-Layer Surfaces; Spherical Harmonics; Ellipsoidal Harmonics; etc. "The great number of particular cases . . . should make the book valuable to geophysicists and others actively engaged in practical applications of the potential theory," Review of Scientific Instruments. Index. Bibliography. xiii + 469pp. 5⅜ x 8.
S486 Paperbound **$2.25**

THEORETICAL MECHANICS: DYNAMICS OF RIGID BODIES, W. D. MacMillan. Theory of dynamics of a rigid body is developed, using both the geometrical and analytical methods of instruction. Begins with exposition of algebra of vectors, it goes through momentum principles, motion in space, use of differential equations and infinite series to solve more sophisticated dynamics problems. Partial contents: moments of inertia, systems of free particles, motion parallel to a fixed plane, rolling motion, method of periodic solutions, much more. 82 figs. 199 problems. Bibliography. Indexes. xii + 476pp. 5⅜ x 8.
S641 Paperbound **$2.00**

MATHEMATICAL FOUNDATIONS OF STATISTICAL MECHANICS, A. I. Khinchin. Offering a precise and rigorous formulation of problems, this book supplies a thorough and up-to-date exposition. It provides analytical tools needed to replace cumbersome concepts, and furnishes for the first time a logical step-by-step introduction to the subject. Partial contents: geometry & kinematics of the phase space, ergodic problem, reduction to theory of probability, application of central limit problem, ideal monatomic gas, foundation of thermo-dynamics, dispersion and distribution of sum functions. Key to notations. Index. viii + 179pp. 5⅜ x 8.
S147 Paperbound **$1.35**

ELEMENTARY PRINCIPLES IN STATISTICAL MECHANICS, J. W. Gibbs. Last work of the great Yale mathematical physicist, still one of the most fundamental treatments available for advanced students and workers in the field. Covers the basic principle of conservation of probability of phase, theory of errors in the calculated phases of a system, the contributions of Clausius, Maxwell, Boltzmann, and Gibbs himself, and much more. Includes valuable comparison of statistical mechanics with thermodynamics: Carnot's cycle, mechanical definitions of entropy, etc. xvi + 208pp. 5⅜ x 8.
S707 Paperbound **$1.45**

THE DYNAMICS OF PARTICLES AND OF RIGID, ELASTIC, AND FLUID BODIES; BEING LECTURES ON MATHEMATICAL PHYSICS, A. G. Webster. The reissuing of this classic fills the need for a comprehensive work on dynamics. A wide range of topics is covered in unusually great depth, applying ordinary and partial differential equations. Part I considers laws of motion and methods applicable to systems of all sorts; oscillation, resonance, cyclic systems, etc. Part 2 is a detailed study of the dynamics of rigid bodies. Part 3 introduces the theory of potential; stress and strain, Newtonian potential functions, gyrostatics, wave and vortex motion, etc. Further contents: Kinematics of a point; Lagrange's equations; Hamilton's principle; Systems of vectors; Statics and dynamics of deformable bodies; much more, not easily found together in one volume. Unabridged reprinting of 2nd edition. 20 pages of notes on differential equations and the higher analysis. 203 illustrations. Selected bibliography. Index. xi + 588pp. 5⅜ x 8.
S522 Paperbound **$2.35**

A TREATISE ON DYNAMICS OF A PARTICLE, E. J. Routh. Elementary text on dynamics for beginning mathematics or physics student. Unusually detailed treatment from elementary definitions to motion in 3 dimensions, emphasizing concrete aspects. Much unique material important in recent applications. Covers impulsive forces, rectilinear and constrained motion in 2 dimensions, harmonic and parabolic motion, degrees of freedom, closed orbits, the conical pendulum, the principle of least action, Jacobi's method, and much more. Index. 559 problems, many fully worked out, incorporated into text. xiii + 418pp. 5⅜ x 8.
S696 Paperbound **$2.25**

DYNAMICS OF A SYSTEM OF RIGID BODIES (Elementary Section), E. J. Routh. Revised 7th edition of this standard reference. This volume covers the dynamical principles of the subject, and its more elementary applications: finding moments of inertia by integration, foci of inertia, d'Alembert's principle, impulsive forces, motion in 2 and 3 dimensions, Lagrange's equations, relative indicatrix, Euler's theorem, large tautochronous motions, etc. Index. 55 figures. Scores of problems. xv + 443pp. 5⅜ x 8.
S664 Paperbound **$2.35**

DYNAMICS OF A SYSTEM OF RIGID BODIES (Advanced Section), E. J. Routh. Revised 6th edition of a classic reference aid. Much of its material remains unique. Partial contents: moving axes, relative motion, oscillations about equilibrium, motion. Motion of a body under no forces, any forces. Nature of motion given by linear equations and conditions of stability. Free, forced vibrations, constants of integration, calculus of finite differences, variations, precession and nutation, motion of the moon, motion of string, chain, membranes. 64 figures. 498pp. 5⅜ x 8.
S229 Paperbound **$2.35**

DYNAMICAL THEORY OF GASES, James Jeans. Divided into mathematical and physical chapters for the convenience of those not expert in mathematics, this volume discusses the mathematical theory of gas in a steady state, thermodynamics, Boltzmann and Maxwell, kinetic theory, quantum theory, exponentials, etc. 4th enlarged edition, with new material on quantum theory, quantum dynamics, etc. Indexes. 28 figures. 444pp. 6⅛ x 9¼.
S136 Paperbound **$2.45**

FOUNDATIONS OF POTENTIAL THEORY, O. D. Kellogg. Based on courses given at Harvard this is suitable for both advanced and beginning mathematicians. Proofs are rigorous, and much material not generally avialable elsewhere is included. Partial contents: forces of gravity, fields of force, divergence theorem, properties of Newtonian potentials at points of free space, potentials as solutions of Laplace's equations, harmonic functions, electrostatics, electric images, logarithmic potential, etc. One of Grundlehren Series. ix + 384pp. 5⅜ x 8.
S144 Paperbound **$1.98**

THERMODYNAMICS, Enrico Fermi. Unabridged reproduction of 1937 edition. Elementary in treatment; remarkable for clarity, organization. Requires no knowledge of advanced math beyond calculus, only familiarity with fundamentals of thermometry, calorimetry. Partial Contents: Thermodynamic systems; First & Second laws of thermodynamics; Entropy; Thermodynamic potentials: phase rule, reversible electric cell; Gaseous reactions: van't Hoff reaction box, principle of LeChatelier; Thermodynamics of dilute solutions: osmotic & vapor pressures, boiling & freezing points; Entropy constant. Index. 25 problems. 24 illustrations. x + 160pp. 5⅜ x 8
S361 Paperbound **$1.75**

THE THERMODYNAMICS OF ELECTRICAL PHENOMENA IN METALS and A CONDENSED COLLECTION OF THERMODYNAMIC FORMULAS, P. W. Bridgman. Major work by the Nobel Prizewinner: stimulating conceptual introduction to aspects of the electron theory of metals, giving an intuitive understanding of fundamental relationships concealed by the formal systems of Onsager and others. Elementary mathematical formulations show clearly the fundamental thermodynamical relationships of the electric field, and a complete phenomenological theory of metals is created. This is the work in which Bridgman announced his famous "thermomotive force" and his distinction between "driving" and "working" electromotive force. We have added in this Dover edition the author's long unavailable tables of thermodynamic formulas, extremely valuable for the speed of reference they allow. Two works bound as one. Index. 33 figures. Bibliography. xviii + 256pp. 5⅜ x 8. S723 Paperbound **$1.65**